Computational Molecular Science and Engineering Forum 2020

Held at the 2020 AIChE Annual Meeting

Online
16 – 20 November 2020

ISBN: 978-1-7138-2314-8

Printed from e-media with permission by:

Curran Associates, Inc.
57 Morehouse Lane
Red Hook, NY 12571

Some format issues inherent in the e-media version may also appear in this print version.

Printed with permission by Curran Associates, Inc. (2021)

AIChE
120 Wall Street, FL 23
New York, NY 10005-4020

Phone: (800) 242-4363
Fax: (203) 775-5177

www.aiche.org

Additional copies of this publication are available from:

Curran Associates, Inc.
57 Morehouse Lane
Red Hook, NY 12571 USA
Phone: 845-758-0400
Fax: 845-758-2633
Email: curran@proceedings.com
Web: www.proceedings.com

TABLE OF CONTENTS

595763 Design of Pore Wall Chemistry to Control Solute Transport and Selectivity

Abstract:

Developing next-generation membranes for water purification demands system-specific materials with precisely tuned functionality. Advances in synthesis have led to the fabrication of isoporous membrane materials with functionalizable pore walls. However, a fundamental understanding of how the chemistry of the wall affects the transport of water and solutes remains elusive, and current practices for avoiding pernicious issues such as fouling largely rely on existing heuristics and trial-and-error methods. Simulations enable the nanoscopic investigation of the mechanism by which pore wall chemistry controls its hydration water and thus its water-mediated interactions with solute molecules. Here, we use molecular dynamics simulations to investigate water and solute structure, dynamics, and transport in a cylindrical nanopore patterned with nonpolar methyl and polar hydroxyl groups. We find that the pore wall chemistry induces structural changes in water near the wall, which lead to differences in macroscopic transport metrics that persist even when the pore is large enough for water in the center of the pore to demonstrate bulk-like behavior. These surface effects furthermore have a significant impact on solute transport through the pore, suggesting that it is possible to achieve solute selectivity via wall functionalization. We thus apply an optimization procedure based on a genetic algorithm to design the chemical patterning of the pore wall to maximize selectivity for a given solute. This inverse design procedure and the accompanying molecular-level understanding of the effect of chemical heterogeneity on solute selectivity in a pore will inform the design of more efficient water purification membranes, as well as the modulation of water-mediated surface-molecule interactions in other systems.

610367 Molecular Engineering of Thixotropic, Sprayable Fluids with Yield Stress Using Associating Polysaccharides

Abstract:

Molecular engineering facilitates the development of a complex fluid with contradictory requirements of yield stress and sprayability, while minimizing the amount of structuring material. Blends of xanthan (Xg) and konjac glucomannan (Kg) containing 99.95 wt% water are demonstrated to satisfy these contradictory requirements by building structure without substantial molecular elasticity. Comparison of shear and extensional properties against a reference solution of polyethylene oxide (PEO), a well-known, Boger fluid, highlights the role of molecular elasticity in controlling critical rheological properties. Added KCl alters inter- and intra-molecular interactions so as to tune rheological properties. The gum blends exhibit a lower extensional resistance with added KCl, which leads to good spray characteristics in contrast to PEO, which shows strain hardening. These measurements suggest that the inter-molecular attractions between the two gums leading to network formation with appropriate stiffness and low molecular elasticity are critical molecular design parameters necessary to achieve sprayable, yields stress fluids.

601186 Electrostatic Potential of Ionic Liquids As an Effective Screening Approach for CO_2 Capture

Abstract:

In response to the need to improve energy efficiency in separations processes such as CO_2 capture from combustion point sources, natural gas sweetening, syngas processing and air separation, several polymers, inorganic and hybrid materials have been developed in recent years as advanced membranes. In the past several years, ionic liquids (ILs) have emerged as potential candidates for CO_2 capture and separation due to their tunability,[1-2] but there is now more interest in designing composite materials that can circumvent some of the common drawbacks of ILs. Accordingly, a better understanding of structure-property relationship of ionic liquids for CO_2 capture is essential for designing new task-specific ILs for future use in composite membrane materials.[3]

Our work aims to provide rational design guidance and structure-property relationships for screening industrial solvents and polymeric membranes based on IL functionalities. It is well-known that quantum mechanical calculation can be used to theoretically study the structure, properties and reaction mechanisms of ionic liquids, but length and time scales are very limited. For decades, it has been recognized that electrostatic potential calculations can provide a great deal of insight for understanding and predicting molecular properties, and this information can be extrapolated from relatively small systems to bulk phase properties. The structural properties (density, molar volume, surface area) and electrostatic potential (ESP) of cations, anions, and combined ion pairs can be generated from first-principles density functional theory (DFT) calculations, as implemented in Gaussian 09,[4] followed by further analysis by the Multiwfn[5] program. Using small DFT-based models, CO_2 molecular probes can be used to sample interactions with IL ion pairs. This provides rapid information about the interaction of CO_2 with ILs that can be used to predict the Gibbs solvation energy and selective adsorption of CO_2 in different IL solvents and IL-based membrane materials. Thus, the structure-property-performance relationships of IL materials for CO_2 capture can be simulated directly from DFT calculations, and the detailed site-site interactions can also be predicted.

Reference:

1. Zhang, X.; Zhang, X.; Dong, H.; Zhao, Z.; Zhang, S.; Huang, Y., Carbon capture with ionic liquids: overview and progress. *Energy Environ. Sci.* **2012,** *5*, 6668-6681.
2. Izgorodina, E. I.; Seeger, Z. L.; Scarborough, D. L. A.; Tan, S. Y. S., Quantum Chemical Methods for the Prediction of Energetic, Physical, and Spectroscopic Properties of Ionic Liquids. *Chem. Rev.* **2017,** *117*, 6696-6754.
3. Bara, J. E.; Carlisle, T. K.; Gabriel, C. J.; Camper, D.; Finotello, A.; Gin, D. L.; Noble, R. D., Guide to CO2 separations in imidazolium-based room-temperature ionic liquids. *Ind. Eng. Chem. Res.* **2009,** *48*, 2739-2751.
4. Frisch, M.; Trucks, G. W.; Schlegel, H. B.; Scuseria, G. E.; Robb, M. A.; Cheeseman, J. R.; Scalmani, G.; Barone, V.; Mennucci, B.; Petersson, G. e., Gaussian~ 09 Revision D. 01. **2014.**
5. Lu, T.; Chen, F., Multiwfn: a multifunctional wavefunction analyzer. *J. Comput. Chem.* **2012,** *33*, 580-592.

610125 Poly(ethylene glycol) Containing Ionic Liquids and Lithium Salt Blends: An Experimental and Molecular Dynamics Simulation Study

Abstract:

PEGylated imidazolium ionic liquids are of interest because of their non-volatility, non-flammability, and unique ion transport properties attributed to poly(ethylene glycol) (PEG). The ether oxygen atoms in the PEG side groups interact with electron deficient centers in the imidazolium ring, leading to increased separation of cation–anion pairs in the fluid, higher ionic conductivity, and higher than expected fluidity for these relatively high molar mass compounds [1]. However, the addition of lithium salts to these ionic liquids is found to result in a significant increase in viscosity, an increase in the glass transition temperature, and a decrease in ionic conductivity [2].

In this work, we used molecular dynamics (MD) simulations to study the interactions of Li^+ in a blend of 1-(monomethoxy PEG)-3-methylimidazolium bis(trifluoromethanesulfonyl)imide (bistriflamide) ([PEGMeIm][Tf_2N]) ionic liquid and lithium bistriflamide ($LiTf_2N$) salt. Simulations show that Li^+ coordinates not only with the ether oxygen atoms of the PEG side group, but also with the oxygen and nitrogen atoms of the $[Tf_2N]^-$ anion, which can adversely affect Li^+ mobility. This finding is corroborated by the following trend of self-diffusion coefficients obtained from the mean squared displacement data: $[PEGMeIm]^+ > [Tf_2N]^- > Li^+$.

Toward improving the lithium ion mobility, we explored solvating it with tetraethylene glycol dimethyl ether (tetraglyme, G4). The 1-(2-methoxyethyl)-3-methylimidazolium bistriflamide ([mEtMeIm][Tf_2N]) ionic liquid, was blended with 1:1 complex of $LiTf_2N$ and G4 (a solvate ionic liquid, denoted herein by [Li(G4)][Tf_2N]). The addition of [mEtMeIm][Tf_2N] to [Li(G4)][Tf_2N] lowers the viscosity and increases the ionic conductivity of the blends, compared with the ionic conductivity of neat [Li(G4)][Tf_2N] [3]. The blends also exhibit improved thermal stability in thermogravimetry experiments.

The dissolution of unchelated $LiTf_2N$ in [mEtMeIm][Tf_2N] results in a decrease in the ionic conductivity at all temperatures of measurement. However, when $LiTf_2N$ is added as a 1:1 complex with G4, the conductivity is higher (at the same molar concentration of $LiTf_2N$). We also find that the complex mixture, comprising of three different ionic species (the lithium ion, and the ionic liquid cation and anion) and one nonionic compound (tetraglyme), of significantly different chemical structures and polarities, show nearly linear mixing rules for density, viscosity and conductivity data.

In this talk, the experimentally determined diffusion coefficients and other transport properties will be analyzed in the context of the MD derived radial distribution functions and coordination numbers.

References

[1] Ganapatibhotla LV, Zheng J, Roy D, Krishnan S. PEGylated imidazolium ionic liquid electrolytes: thermophysical and electrochemical properties. *Chemistry of Materials.* 2010; 22(23): 6347-6360.

[2] Wu L, Venkatanarayananan RI, Shi X, Roy D, Krishnan S. Glass transition, viscosity, and conductivity correlations in solutions of lithium salts in PEGylated imidazolium ionic liquids. *Journal of Molecular Liquids.* 2014; 198: 398-408.

[3] Wang Y, Turk MC, Sankarasubramanian M, Srivatsa A, Roy D, Krishnan S. Thermophysical and transport properties of blends of an ether-derivatized imidazolium ionic liquid and a Li^+-based solvate ionic liquid. *Journal of Materials Science.* 2017; 52(7): 3719-3740.

609990 The Interplay Morphology and Sequence on Ion-Coupled Electron Transport in Mixed Conducting Polymers

Abstract:

Macromolecules that exhibit both electron transport and ionic mass transport (i.e., mixed conducting polymers) are ascendant with respect to both emerging application spaces and the elucidation of their fundamental physical principles. The unique coupling between the two modes of conduction puts these materials at the center of many next-generation organic electronic applications, including organic electrochemical transistors. However, the degree to which ion motion is coupled with electron transport and morphology changes, and the range of polymer chemistries that can exhibit and be optimized for electrochemical response in aqueous media is unknown. To elucidate the general relationship between polymer structure, morphology, and mixed conduction we have developed a hybrid simulation methodology consisting of coarse-grained modeling of mixed conducting materials and master-equation modeling of electron transport to characterize long-timescale morphology reorganizations and its impact on mixed conduction. Using this method, we have performed a systematic study of several polymer design components, including side-chain placement and constitution, repeat unit sequence, and tailored π-π interactions to develop generic design rules for next-generation mixed conducting polymers. The results highlight the critical role of electron-ionic coupling in modulating the conduction of these materials and suggests the contemporary modular design paradigm of separately optimizing electron and ion conducting components leaves out this critical physics.

605784 Building Block and Self-Assembly Design with Magnetic Handshake Materials

Abstract:

Inspired by nature's ability of self-assembling intricate materials, we recently developed a self-assembly platform that can create specific binding by encoding magnetic dipole patterns into panel-like building blocks. With this platform, which we term magnetic handshake materials, we are able to achieve controlled polymerization, complementary binding strands and 3D folding from 2D nets. However, in order to perform more systematic materials design that can match nature's ability, we need to develop a design framework that can not only search through the high dimensional parameter space more efficiently, but also define the design limits of our magnetic building blocks. Here, I will present a design protocol inspired by information theory that can provide us the upper limit of the number of distinct building blocks for a given design complexity, and then show how to use these building blocks to self-assemble structures with non-trivial geometry and topology.

610576 Self-Assembly in Size-Disperse Binary Systems

Abstract:

Binary systems of nanoparticles are promising targets for self-assembled materials due to their great potential for structural diversity (compared with one-component systems). Hard particle models often fail to capture the phase behavior of binary systems [1]. Previously, the effect of particle size dispersity on structural stability has been examined almost exclusively in nanosphere and disk phases, with only a few studies investigating particles with, *e.g.*, diverse anisotropic properties [2,3]. We investigate molecular dynamics simulations of binary systems of particles interacting *via* attractive isotropic potentials that assemble structures of various symmetries and complexities. We report different crystal phases that are more or less sensitive to the introduction of differently-sized particles, and we discuss the use of this study to predict robust systems for improved self-assembly experiments on the mesoscale.

References:

[1] X. Ye, C. Zhu, P. Ercius, S. N. Raja, B. He, M. R. Jones, M. R. Hauwiller, Y. Liu, T. Xu, A. P. Alivisatos, "Structural diversity in binary superlattices self-assembled from polymer-grafted nanocrystals", Nat. Commun. 6, 10052 (2015).

[2] C. L. Phillips, C. R. Iacovella, S. C. Glotzer, "Stability of the double gyroid phase to nanoparticle polydispersity in polymer-tethered nanosphere systems", Soft Matter 6, 1693-1703 (2010).

[3] C. L. Phillips, S. C. Glotzer, "Effect of nanoparticle polydispersity on the self-assembly of polymer tethered nanospheres", J. Chem. Phys. 137, 104901 (2012).

609167 Using Defects to Reshape Colloidal Assemblies

Abstract:

Directed assembly at the colloidal scale remains challenging despite advances in using optical forces to manipulate nanoparticles. Optical holography allows for precise positioning of components, but does not scale well to thousands or millions of particles. However in dense colloidal systems particle positions are highly correlated, and so not every particle need be controlled in order to enact reconfigurations of the dense assembly. In this work we use simulation to study how a small number of particles embedded within a colloidal monolayer can exert forces which lead to long range permanent deformation of the monolayer edges. By framing the problem in terms of the creation and migration of mobile dislocation defects, we uncover design principles for the embedded cluster shape which result in the accumulation of large shear deformations on lattice planes. We show that arbitrary control of shear deformation localized to a single plane is sufficient to carry out a variety of useful reshaping operations on crystalline colloidal monolayers. This work represents a method to increase the length scale of control achieved with optical forces without imposing additional complexity on the optical system required.

607840 Simulation-Driven Design of Chemoresponsive Liquid Crystal Sensors By Tuning Analyte Partitioning

Abstract:

Chemoresponsive liquid crystal (LC) sensors are promising platforms for the detection of vapor-phase analytes. These sensors consist of a LC thin film deposited on a reactive substrate. Upon exposure to an analyte, changes to the anchoring of the LC film to the substrate triggers a change to LC order which is observable optically. Experiments have determined that the activation time of the sensor depends on the transport of analytes within the LC films. Understanding and predicting these transport properties can guide the design of LC sensors with improved selectivity.

In this work, we use atomistic molecular dynamics simulations to quantify the partitioning and diffusion of nine small-molecule analytes, including four common atmospheric pollutants, in model systems representative of LC sensors. We first parameterize all-atom models for 4-cyano-4′-pentylbiphenyl (5CB), a mesogen typically used for LC sensors, and all analytes. We validate these models by reproducing experimentally determined 5CB structural parameters, 5CB diffusivity, and analyte Henry's law constants in 5CB. Using the all-atom models, we calculate analyte solvation free energies and diffusivities in bulk 5CB. These simulation-derived quantities are then used to parameterize a mass-transport model to predict sensor activation times. These results demonstrate that differences in analyte–LC interactions can translate into distinct activation times to distinguish activation by different analytes. Finally, we quantify the effect of LC composition by calculating analyte solvation free energies in TL205, a proprietary LC mixture. These calculations indicate that varying the LC composition can modulate activation times to further improve sensor selectivity. These results thus provide a computational framework for guiding LC sensor design by using molecular simulations to predict analyte transport as a function of LC composition.

608021 Polymer Informatics: Current Status & Critical Next Steps

Abstract:

The launching of Materials Genome Initiative (MGI) in 2011 has marked a milestone of materials informatics, after which materials design underwent a paradigm transform. Among notable applications within this research and development area, the successful deployment of novel materials has been greatly accelerated by the use of computational, experimental and informatics methodologies. Here, we describe our recent polymer discovery efforts, highlighting the role played by computational data generation and screening, targeted synthesis and characterization, polymer fingerprinting and machine-learning prediction models. These efforts have culminated in the creation of an online Polymer Informatics platform (https://www.polymergenome.org) to guide ongoing and future polymer discovery and design. Challenges that remain are examined, and systematic steps that may be taken to extend the applicability of such informatics efforts to a wide range of technological domains are discussed. This work is supported by Office of Naval Research.

605670 Using Derivative Information from Molecular Simulations to Enhance Predictions of Structural Properties across State Conditions

Abstract:

Accurately determining shifts in structural properties is crucial for predicting phase and self-assembly behavior in molecular fluids. In the case of water, for example, such transitions provide a basis for defining regions of anomalous thermodynamic and dynamic properties. Thermodynamic extrapolation has previously been shown to provide information about arbitrary structural observables at temperatures (or relative chemical potentials in open-system mixtures) different from those at which simulations were performed. This relies on the use of derivative information that is readily available from molecular simulations, though infrequently utilized. We present a detailed analysis of the uncertainty associated with derivatives of structural properties with respect to temperature and density for both an analytically solvable ideal gas model, as well as rigid water. We present formulae for extrapolating in volume for canonical ensembles and demonstrate that linear extrapolations of water's structural properties are only accurate over a limited density range. On the other hand, linear extrapolation in temperature can be accurate across the entire liquid state. We demonstrate conditions under which extrapolation methods are applicable, clearly explaining performance through connections to re-weighting techniques associated with free energy calculations. Through such considerations, we propose a recursive algorithm implementing polynomial interpolation to efficiently predict variation in structural properties with state conditions, as well as a simple visual self-consistency check to facilitate its use.

608221 Using Advanced Sampling Techniques to Study Solute and Ion Transport through Semipermeable Membranes

Abstract:

The ability of semipermeable membranes to selectively impede the transport of undesirable ions and molecules is key to many applications, from desalination and gas separation, to biological membrane transport. For instance, membranes that are only permeable to water and reject most other ions and molecules are used in water desalination. Designing more selective membranes requires characterizing the kinetics and mechanism of the transport of unwanted entities. The timescales associated with such processes, however, can be too long to be accessible to conventional molecular dynamics (MD) simulations. Moreover, the separation of timescales between the transport of desirable entities (such as water) and undesirable entities (such as salt) will result in considerable changes in concentration throughout an MD trajectory, which can skew the kinetics and mechanism of transport in ways that are difficult to correct. Recently, we utilized jumpy forward flux sampling (jFFS) [1] and non-equilibrium MD to study pressure-driven ion transport through graphitic nanoporous membranes [2]. This approach not only allows us to accurately estimate arbitrarily long mean passage times, but also to compute fluxes and permeabilities within a pseudo-equilibrium ensemble in which both reservoirs are at different— but almost constant— chemical potentials. In this presentation, we will discuss the technical aspects of this approach in the context of a few more examples, with the goal of addressing interesting questions about ion transport and selectivity.

[1] Haji-Akbari, J. Chem. Phys., 149: 072303 (2018).

[2] Malmir, et al., Matter, 2: 735 (2020).

599700 An Efficient Method to Calculate the Diffusivity of Non-Spherical Molecules in Nanoporous Materials

Abstract:

Conventional methods for predicting diffusion coefficients in nanoporous materials are computationally expensive and the difficulty increases for non-spherical gas molecules and for materials with extremely small pores. Whereas the problem is partially alleviated by recent development in fast computers, the data-driven approach to materials design often requires a large amount of data over a broad range of conditions. To develop a more efficient method to calculate the diffusion coefficient, we extend the sting method for predicting the minimum free energy pathway (MFEP) by introducing collective variables applicable to non-spherical molecules. The combination of MFEP with the transition-state theory (TST) allows us to predict diffusion coefficients in good agreement with molecular dynamics simulation but with much less computational cost. The new theoretical procedure is promising for applications to data-driven approach to inverse design of nanoporous materials for membrane-based separation.

611267 Investigating Path Sampling Techniques on Complex Energy Landscapes to Study Nucleation

Abstract:

Nucleation is a rare event in molecular simulations and thus, requires enhanced sampling techniques. One category of these are path sampling techniques of which forward flux sampling (FFS) and transition interface sampling (TIS) have been used extensively to study nucleation. These methods have several advantages – they do not bias the dynamics of the systems, they are not dependent on the order parameter, and result in an ensemble of transition paths enabling detailed mechanistic studies of the transition. There are, however, some practical challenges in applying these techniques for complex energy landscapes, which direct affect the efficiency of these methods. In our study we have compared FFS and TIS for nucleation in Lennard Jones systems, which represent a complex energy landscape. We provide in-depth discussion of the methodological details important to ensure the appropriate implementation of these methods. Further, we evaluate the efficiencies of the methods and comment on their relative strengths and weaknesses in context of complex energy landscapes. This study fills a knowledge gap in the field – while many methods are available, and most of them are computationally expensive, it is non-trivial to compare them. Our studies will thus provide insights and guidelines for the selection of appropriate method for rare event simulations to the scientific community.

609325 Coupling Enhanced Sampling with Monte Carlo Techniques Improves Flexible-Backbone Protein Docking

Abstract:

Protein-protein interactions (PPIs) govern nearly all biological mechanisms, ranging from enzyme catalysis and inhibition to signaling and gene regulation. Understanding protein function and the associated dynamics of protein interactions at the molecular scale is key in delineating disease mechanisms, including COVID-19, cystic fibrosis, Alzheimer's and cancer. Current computational tools are often confounded by the many degrees of freedom for structural flexibility of protein backbones, thereby hindering the accuracy of protein complex structure predictions. To overcome this challenge and to mimic binding-induced conformational changes in protein complexes, we integrated enhanced sampling methods, including temperature and Hamiltonian replica exchange, with docking protocols in Rosetta to improve local and global docking predictions. Our method allows us to rapidly explore relative orientations between protein partners with enhanced sampling. To capture backbone degrees of freedom, we sample backbone motions on residues comprising a putative interface on-the-fly, thereby incorporating binding-induced changes with docking. We benchmarked our protocol on 20 protein complexes starting from unbound binding partners. In a global search, i.e. without any prior knowledge of the binding site, we are able to reach lower energies and improve native structure discrimination compared to RosettaDock4 and ReplicaDock. With 2 x 10^6 MC steps employed for each target, we were able to identify near-native geometries using 300-500 CPU-hours, depending on the protein sizes, which is relatively efficient compared to conventional molecular dynamics. In local coarse-grained docking, our method obtains sub-angstrom predictions for about half of the benchmark targets. This approach successfully samples global docking energy landscapes in rigid-body coordinates, and it captures several Angstroms of induced backbone conformational changes at the protein-protein interface.

598044 Computational Reverse-Engineering Analysis for Scattering Experiments of Assembled Binary Colloidal Particle Mixtures

Abstract:

Directed and self-assembly of nanoparticle/colloidal mixtures is a proven route to precisely engineer organic/inorganic materials with controlled optical properties. During the design of such optically active materials, an important step is the structural characterization of the assembled colloid/particle mixture. In many cases, researchers use small angle neutron or X-ray scattering to conduct such structural characterization. The intensity profiles, I(q) vs. q, that are produced from scattering experiments are then interpreted to real space arrangement using off-the shelf analytical models that can sometimes be too approximate for the system at hand. In this talk, we will present our recent work in development of computational methods to analyze the structure of binary nanoparticles assembled into supraballs from the intensity profiles obtained by contrast-matched small angle neutron scattering (SANS). Given the contrast-matched scattering intensity profiles and information about the nanoparticle size and dispersity as inputs, our method utilizes a combination of a genetic algorithm and simulations to output the best structure of the self-assembled nanoparticles whose computed intensity profile matches the experimental intensity profile. This method provides nanoparticle level structural information in complex assembled geometries such as spherical confinements, without dependence on off-the-shelf analytical models.

599612 Discovery of Light-Driven Hydrogen Evolution Catalysts Using High Throughput Simulation, Experiments and Machine Learning

Abstract:

The renewable production of hydrogen is an important technology. Current approaches to catalyst discovery for thermo/electro and photochemical methods for hydrogen production have been difficult to optimize. This difficulty arises in part because of the separation of the synthesis and characterization of catalysts from their performance under operating conditions. It is challenging to build models that relate catalyst performance to synthesis conditions. We have taken a different approach to this problem. We use a photo-driven process that couples a photoabsorber with the synthesis of multi-component metal particles and hydrogen production. This approach is readily parallelized in a multi-well plate, enabling us to explore a wide range of synthesis conditions that are directly connected to catalyst performance measurements. We will show how we used this approach to screen thousands of possible catalyst synthesis conditions to identify promising combinations. The high-throughput experiments along, however, do not tell us what metals to test. For this we augment the approach with a parallel high-throughput computation approach where hundreds of thousands of DFT calculations are performed on multi-component alloy surfaces to screen for promising candidates with desired hydrogen and CO adsorption energies. Finally, we combine these efforts to drive focused studies of the most promising candidate synthesis conditions to develop a more conventional understanding of how the synthesis conditions led to the observed activity. We will show how this approach has led to confirmation of some known results from the electrochemical production of hydrogen from water, as well as some new results we may not have discovered with any single approach.

601035 High Throughput Screening of Alloy Structures for Propane Dehydrogenation Reaction

Abstract:

Non-oxidative propane dehydrogenation (PDH) on intermetallic alloys is a promising technology to convert large amounts of propane available in the shale gas to propylene. However, the high temperatures required for the reaction lead to reduced selectivity towards propylene formation and deactivation of the catalyst surface.[1] Even though Pt and Pd alloys have shown promise in terms of improved selectivity in comparison to pure metals, they ultimately experience loss of reactivity under reaction conditions. This difficulty motivates development of a more fundamental understanding of why Pt and Pd alloys have improved performance compared to pure metals and, further, elucidation of how such insights can be leveraged to find other alloy combinations with simultaneously high selectivity, stability, and activity for PDH.

Through our previous analysis on PdIn alloys, we have demonstrated that, as compared to terraces, the step surfaces of 1:1 alloys are highly active under PDH conditions (5 orders of magnitude higher rates). Furthermore, a comprehensive microkinetic analysis pointed to useful descriptors for activity, selectivity and stability.[2] With these descriptors as a starting point, the current study focuses on high-throughput screening of intermetallic alloy structures, including both steps and terraces, to identify reactivity trends across Pt and Pd alloys for PDH and, ultimately, find improved catalysts for this chemistry. First, we deploy a simple automated framework for adsorption and activation energy calculations on large number of chosen terrace and step alloy structures. The Materials Project is used to identify the most stable bulk structure of the alloy composition, while a surface energy analysis points to the most stable surface terminations, and CatKit is employed to identify all the distinct adsorption sites [3]. Subsequently, an in-house algorithm identifies all unique adsorbate configurations and finds the most stable adsorption site. These data are organized with a python-based databasing tool, leading to identification of BEP-type correlations between bond breaking barriers and the binding energies of various adsorbates. Further the binding energies of key descriptors were correlated with electronic and geometric properties of the alloy surfaces, using Random Forest techniques. This, therefore, allowed for a deeper understanding of the surface features that impacts the binding of descriptors the most.

References:

[1] J.J.H.B. Sattler, J. Ruiz-Martinez, E. Santillan-Jimenez, B.M. Weckhuysen, Catalytic dehydrogenation of light alkanes on metals and metal oxides, Chem. Rev. 114 (2014) 10613–10653. https://doi.org/10.1021/cr5002436.

[2] R.R. Seemakurthi, et al. Structure Sensitivity and Microkinetic Analysis of Propane Dehydrogenation on PdIn alloy and Pd, In Preparation.

[3] J.R. Boes, O. Mamun, K. Winther, T. Bligaard, Graph Theory Approach to High-Throughput Surface Adsorption Structure Generation, J. Phys. Chem. A. 123 (2019) 2281–2285. https://doi.org/10.1021/acs.jpca.9b00311.

609929 Theory-Guided, Interpretable Machine Learning Finds Predictive Geometric Structure-Property Relationships for Chemisorption on Alloys

Abstract:

Chemisorption, the direct chemical bonding of an atomic or molecular species to a solid-state material, is central in the fields of catalysis, corrosion, and electrochemistry, among many others. Despite the fundamental nature and importance of chemisorption, linking the geometry (i.e., the structure and composition) of different materials to their chemisorption properties remains a critical challenge. Developing physically transparent and quantitatively accurate models that can relate the chemisorption strength between an adsorbate and a solid surface to the adsorption site's geometry is critical to advance our understanding of chemisorption. In this talk, we discuss our efforts to use a theory-guided machine learning approach, which uses an interpretable class of machine learning models called generalized additive models (iGAM models),[1] to discover predictive structure-property models that can quantify the chemisorption strength of O, OH, S, and Cl on Pt-metal and Au-metal alloy surfaces subject to various strain- and ligand-induced changes in the local geometric structure. The iGAM models show a strong degree of predictive accuracy, with an average root-mean-square-error (RMSE) of 0.046 eV for samples in the test set. Through quantification of the relative importance of the features used to construct the models, we identified three important geometric features of the adsorption site that impact the relative chemisorption strength on metal alloys: the strain in the surface layer, the number of *d*-electrons in the ligand metal, and the size of the ligand atom. The interpretable functional form of the iGAM models allows us to analyze the model behavior with respect to each of these geometric features. Comparison between the iGAM chemisorption models and established electronic-structure models shed light on the critical physical concepts that control the chemisorption process on metal surfaces.

References

(1) Lou, Y.; Caruana, R.; Gehrke, J. Intelligible Models for Classification and Regression. In *Proceedings of the 18th ACM SIGKDD International Conference on Knowledge Discovery and Data Mining*; KDD '12; ACM: New York, NY, USA, 2012; pp 150–158. https://doi.org/10.1145/2339530.2339556.

608786 Trends in Catalyst Stability and Reactivity: Extracting Physical Insights Using Simple Data-Driven Approaches

Abstract:

The catalytic turnovers of a typical reaction pathway are determined by the energies of between 10 to 100 reaction intermediates.[1] Discovering optimally performing catalysts entails maximizing catalytic turnovers across vast spaces of energy and materials. One way to simplify this formidable optimization problem is to parametrize the reaction mechanism using energy correlations.[2] These energy correlations, often linear functions, are mappings between binding energies of reaction intermediates (e.g. $C_xH_yO_z*$) and one to two catalytic descriptors (e.g. binding energies of CH* and O*). Such energy correlations have enabled the high-throughput screening of metals, oxides, and zeolites. The slopes of these linear correlations (e.g. 0.5 for correlations between OH* and O*) across metals, oxides, and zeolites are determined by the ratios of bond orders of the corresponding species. Recent advances notwithstanding, these correlations are limited by three key features. First, since these correlations do not include catalyst stability metrics, they are unable to describe catalyst dynamics under reaction conditions. Second, within the existing formalism it is difficult to reverse-engineer active sites through property (e.g. rate) → structure (e.g. chemical composition) "inverse design" approaches. Third, their linearity, together with rigid constraints on the slope, imposes severe limitations in the accessible regions of materials space. There has been a concerted effort in the computational catalysis community to bridge these three knowledge gaps using data-driven statistical learning models.[3]-[5] Herein, we illustrate our contributions towards reverse-engineering realistic active ensembles sites that possess targeted catalytic turnovers.[6]-[9] We place a special emphasis on extracting physical insights from regressed parameters within our simple data-driven approach.[6], [10]

In the first part, we present accelerated schemes for determining the energies of metal atoms that, in turn, influence catalyst stability.[6] Predicting energies of metal atoms with atomic site specificity enables expeditious evaluations of surface energies, cohesive energies of nanoparticles having generic shapes, and binding energies of metal adsorption sites. We calculate the energies of metal atoms by melding quadratic and linear interpolations across the space of coordination numbers, composition, and external strain.[7] These new interpolations shrink the training set size for 12 fcc p- and d-block metals from > 204 to as few as 24 Density Functional Theory (DFT) calculations, without sacrificing accuracy on the test set.[6] We will specifically highlight physical insights regarding metal-metal bonding that are obtained from the regressed parameters. As a proof of concept, we employ this method to reverse engineer thermodynamically stable active site motifs of nanoparticles that catalyze C_1 conversion.

In the second part, we investigate the existence of energy correlations on supported nanoparticles using Au nanorods on doped MgO supports as a prototypical model.[10] Using > 1700 DFT calculated binding energies, we show that although DFT-derived energy correlations between 31 reaction intermediates of CO oxidation, water gas shift, and methanol synthesis do exist, their slopes depart from bond order conservation constraints. This departure from bond order conservation at bifunctional metal/oxide interfaces enhances the flexibility in materials space during catalyst optimization. We build a conceptual framework using the multipole expansion in electrostatics to semi-quantitatively explain trends within the slopes. Through these trends, we elucidate the electrostatic origin of the departure in slopes from bond order conservation.

Taken together, these two studies reveal new functional forms for energy correlations that determine metal nanoparticle stabilities, and reactivity trends at bifunctional interfaces. More crucially, we show that parameters of simple data-driven models contain a wealth of intriguing physical insights that enhance model interpretability.

References

1. Norskov et al., *Proc. Natal. Acad. Sci. USA,* (2011), **108,** 937.
2. Abild-Pedersen et al., *Phys. Rev. Lett.* (2007), **99,** 016105.
3. Tran et al., *Nature Catal.* (2018), **1**, 696.
4. O'Connor et al., *Nature Catal.* (2018), **1**, 531.
5. Goldsmith et al., *AICHE J.* (2018), **64,** 2311.
6. Choksi, Streibel, and Abild-Pedersen., *J. Chem. Phys.* (2020), **152,** 094702.
7. Streibel, Choksi and Abild-Pedersen. *J. Chem. Phys.* (2020), **152,** 094701.
8. Choksi et al. *J. Phys. Chem. Lett.* (2019), **10,** 1852.
9. Roling, Choksi, and Abild-Pedersen. *Nanoscale* (2019), **11,** 4438.
10. Choksi et al. *Angew. Chem. Int. Ed.* (2018), **57,** 15410.

604578 Orienteering in an Uncharted Chemical Space: Searching for an Optimal Bimetallic Nanocatalyst

Abstract:

Nanoparticles (NPs) have received tremendous attention as catalysts due to their high specific surface area and abundance of unique active sites. When designing such a catalyst, the Sabatier principle tells us that it must be tuned to adsorb key intermediates neither too strongly nor too weakly. For this reason, it is imperative to screen candidate NPs for their adsorption behavior. This task is conventionally accomplished either through time-consuming experiments, or computationally-expensive *ab-initio* calculations, which are primarily limited to highly idealized periodic surfaces. This presents a bottleneck in materials discovery, one which can be overcome through the development of rapid methods to predict adsorption behavior. To this end, we present a universal model of small molecule adsorption to NPs [1], which is able to accurately capture the adsorption behavior of metal NPs of any size, shape, or composition with minimal computational cost. By combining this adsorption model with our new genetic algorithm approach to NP structure prediction, we are now able to efficiently traverse the vast chemical space of bimetallic NPs and identify possible catalytic NPs to feed experiments. Overall, this work leverages cutting-edge chemical modeling techniques, such as first principles calculations, machine learning and statistical regression to significantly accelerate materials discovery and enable the rapid identification of promising bimetallic nanocatalysts.

1. Dean, J.; Taylor, M.; Mpourmpakis, G. Unfolding Adsorption on Metal Nanoparticles: Connecting Stability with Catalysis. *Science Advances* 5 (9), eeax5101.

604822 Physics Informed Machine Learning of Chemisorption at Metal Surfaces

Abstract:

The formation and breakage of chemical bonds at active sites is the molecular basis of catalysis. Being able to rapidly compute interaction strengths between bonding entities and understand their trends holds the key to the design of improved catalysts. Despite recent advances, machine learning (ML) faces a tremendous challenge for catalysis applications due to its poor transferability and explainability. Here we present a physics informed machine learning (PIML) approach that integrates convolutional neural networks with the *d*-band theory of chemisorption for predicting the chemical reactivity of metal surfaces. With *OH and *CO as two representative adsorbates, we demonstrated that the hybrid ML models outperform the purely data-driven ones in both data scarce and rich regions, especially for out-of-sample systems. More importantly, the architecture design enables its physical interpretability, shedding light on the nature of chemical bonding at metal surfaces.

606955 Improved Catalyst Predictions with Machine Learning Coupled Alchemical Perturbation Density Functional Theory

Abstract:

First principles Kohn-Sham density functional theory (KS-DFT) calculations hinder large-scale searches for new and sustainable heterogeneous catalysts due to significant computational expense. Alchemical perturbation density functional theory (APDFT) is a promising method that provides rapid predictions of catalyst descriptors based on perturbations of electrostatic potentials that arise from isoelectronic transmutations made to a reference catalyst. While straightforward and simple, first-order APDFT approximations are largely inaccurate when large and/or many transmutations are made. Here, we first discuss systematic trends in errors related to these factors for binding energy (BE) prediction of CH_x, NH_x, and OH_x adsorbates on hypothetical alloy variations of fcc Pt(111) surfaces. Based on these observations, we demonstrate our approach fingerprinting locations of transmutations in these hypothetical alloys, and we show how to correct APDFT errors with machine learning models trained using about 3600 BE data points. Our models provide improved prediction accuracy with errors decreased by as much as an order of magnitude, and this approach extends the breadth of APDFT predictions, allowing rapid and accurate BE predictions on many more alloys made by numerous transmutation combinations.

607056 Predictive Design of Biomimetic Nanomaterials Via Bottom up Approaches

Abstract:

Developing transformative approaches for engineering materials and technologies is crucial for addressing urgent national and global challenges in healthcare, renewable energy and the global environment. One promising solution to this growing problem is to develop synthetic materials and devices that can mimic the complex functions of biological matter. Towards this end, bottom up approaches including self and directed assembly techniques have been shown as a powerful means for engineering the underlying nanostructure of this exciting class of materials and devices. The fundamental challenges to these bottom up techniques are to design the optimal nano building blocks such as block copolymers, nanoparticles and colloids, and efficient assembly pathways for desirable nanostructures. In this talk, I will summarize my past and current studies that are devoted to addressing these challenges. I will show how the size and shape of the assembled nanostructures can be controlled through the electrostatic interactions between nanoparticles and colloids. The resulting assemblies are highly uniform size, reminiscent of virus capsids. I will also demonstrate the ability of random copolymers, which mimic intrinsically disordered proteins in living cells, to encapsulate and protect numerous proteins in non-native media. The assembled polymer-protein complexes are designed so that they function as nanoreactors where the proteins inside retain their activities. Also presented are the algorithms and methods I have been developing to improve the efficiency of simulation studies in general, and for my studies in particular, ranging from massively parallel simulation codes to enhanced sampling techniques, which are now available to the public through the software packages LAMMPS and HOOMD-Blue. Overall, the goal of my studies is to accelerate the design of adaptive, programmable nanomaterials for a broad range of applications including, but not limited to, drug delivery and release, biosensing, bioremediation, energy storage and conversion.

609231 A Microscopic Theory of Entropic Bonding for Colloidal Crystal Prediction

Abstract:

Advances in the synthesis of hard polyhedral particles have directed attention towards their usage in self-assembly as they give rise to a wide range of interesting morphologies. Yet no classical model exists for that allows us to predict the assembly behaviors for anisotropic particles. As a result, theoretical descriptions, based purely on a particle's geometry, that can systematically describe the microscopic origins for the directional entropic forces between particles are of significant interest. Here, we present one such theory for hard particle shapes that self-assemble solely through entropy maximization into colloidal crystals, including those of remarkable complexity. This microscopic theory of entropic bonding employs the use of shape orbitals to enable *a priori* prediction of thermodynamically preferred colloidal crystal structures, in good agreement with previously published simulations across multiple systems.

609194 Dynamic Materials from Colloidal Crystals: Active Matter Coupled to Crystalline Defects

Abstract:

When many thousands or millions of designed components are combined a new class of material is formed, generally known as a metamaterial. When the subcomponents are capable of exerting individual forces, sensing, or communicating a truly exotic material with potentially life-like qualities is produced. The design and control of such a material for engineering purposes is a daunting task, especially when the size of each subcomponent is desired to be very small (i.e. sub-micron); In such a case the sensing, actuation, and communication abilities of each subunit are severely restricted.

We present a strategy for controlling the macroscopic properties of artificial crystalline materials by taking control over the dynamics of dislocation lattice defects. In this scheme, the majority of the material is passive, with a small fraction of active components capable of exerting individual forces. By coupling the active components to lattice defects, their range of influence can be extended far enough to control macroscopic properties.

Using this concept, we demonstrate (via simulation) how bulk material properties such as resistance to plastic deformation can be controlled, as well as how directed dislocation motion can be used for actuation and re-shaping of crystalline metamaterials. We show how the strain field of mobile active interstitials within the material can be tuned via Monte Carlo optimization for maximized coupling, as well as how forces generated from an embedded cluster of active material can be designed to create dislocation pairs which can be used for bulk material reconfiguration.

By only using a small fraction of active components, the planning, control, and component complexity requirements necessary to design the collective behavior of the material are relaxed, paving the way for experimental realization of adaptive active metamaterials at the colloidal scale.

598066 Liquid-Liquid Phase Transition in Water from a Machine-Learning First Principles Force Field

Abstract:

Finding an explanation for water's anomalous thermophysical properties (*e.g.*, liquid water's observed maximum in density, and sharp increases in compressibility, heat capacity, and the magnitude of the thermal expansion coefficient upon cooling) is an area of vigorous research activity and debate. One theory is that under strongly supercooled conditions and high pressures, water undergoes a liquid-liquid phase separation into metastable high-density and low-density liquid states, and the thermodynamic implications of this transition can be traced to explain some of water's anomalies. Computer simulation techniques have led the charge in studying this phenomenon, as they are not in general affected by rapid crystallization, the occurrence of which in experiments places the temperatures and pressures where the liquid-liquid transition is theorized to occur outside the range accessible by current experimental techniques. As such, the search for ever more advanced computational approaches to definitively establish (or disprove) water's liquid-liquid transition is an active area of research. Herein, we use the Deep Potential Molecular Dynamics (DPMD) technique to generate a neural network-based model of water obtained from high-accuracy density functional theory calculations. We then use biased DPMD simulations in the multithermal-multibaric ensemble to efficiently obtain the thermophysical properties of this model over a wide range of temperatures and pressures. We analyze these results in the context of a phenomenological two-state equation of state, and we discuss how our findings provide *ab-initio* evidence consistent with a liquid-liquid transition. Lastly, we describe ongoing efforts to apply free energy techniques to definitely establish liquid-liquid coexistence using this DPMD model. This work brings state-of-the-art machine learning, *ab-initio*, and statistical mechanical methods to significantly deepen our understanding of the sources of water's unusual properties, bolstering the hypothesis according to which this most ubiquitous yet unusual substance possesses two critical points.

599406 Development of Coarse-Grained Polymer Models for Chemistries with Hydrogen Bonding Capability

Abstract:

Functional materials engineered with polymer chemistries capable of specific and directional interactions, like hydrogen bonds, exhibit unique structural organization and dynamics. In particular, the morphology (i.e., shape, size and local arrangements of the assembled supramolecular structures) and thermodynamics within these materials can be tuned by tailoring the placement of hydrogen bond acceptor and donor atoms at strategic locations along the polymer chain(s). Even though simulations with atomistic and existing coarse-grained (CG) models have been valuable in understanding or predicting structure within polymeric systems, they either lack the ability to simulate large length and time scales (in atomistic models) or lack the monomer-level directionality imparted by such interactions (in CG models). In this talk, we will present our recent work extending a previously published generic CG polymer model that captures the directionality of hydrogen bonding interactions within polymers [1] to represent two specific polymer chemistries, poly(vinyl-4-phenol) and poly(vinyl-2-pyridine) homopolymers. We will highlight the modifications needed to the generic CG model to reproduce atomistically observed structure and conformations for these two chemistries.

[1] Kulshreshtha, A.; Modica, K. J.; Jayaraman, A. Impact of Hydrogen Bonding Interactions on Graft-Matrix Wetting and Structure in Polymer Nanocomposites, Macromolecules 2019, 52 (7), 2725−2735.

596718 Multiscale Optimization in Biomimetic Assembly

Abstract:

Clarifying the driving mechanisms involved in various biological processes (i.e., protein aggregation, membrane fusion, etc.) has become one of the main goals in our scientific community. One of the promising directions for this task is by performing molecular simulations of analogous biomimetic assembly. In this presentation, I correspondingly discuss the two main research routes which I am pursuing: Particularly, together with my colleagues, I investigate hydration forces between colloids and phase transitions in polymers. First, in the colloidal project, we foremost resolve a paradox that has existed for a half of a century between numerical predictions and empirical measurements regarding the hydrophobic force: We specifically find a continuous transition between the association of small hydrophobes (below ~1 nm in size), which assemble entropically via a weak oscillatory force, and the association of large hydrophobes (above ~1 nm in size), which assemble enthalpically via a strong monotonic force; we explain this length-scale crossover in terms of the ability of a water solvent in forming a tetrahedral network around nonpolar solutes [1-2]. Besides, we also systematically transform these biomimetic colloids into hydrophiles, finding a multifaceted energy-scale crossover for the hydrophilic force [3]. Second, in the polymeric project, we examine a bead-spring chain that embodies a coil-crystal transition: Specifically via perturbation theory, we show that such a biomimetic polymer characterizes the universal kinetics associated with the force spectroscopy of various biological chains [4]. We continue by studying the influence of this coil-crystal transition of a single polymer on the corresponding aggregation propensity of many polymers. On a practical level, our molecular simulations for the biomimetic assembly of colloids and polymers can be of much aid in the design of pharmaceuticals.

[1] A. Chaimovich and M.S. Shell (PRE, 2013).
[2] A. Chaimovich and M.S. Shell (PRE, 2014).
[3] A. Chaimovich (In preparation).
[4] A. Chaimovich (In preparation).

606419 Computational Design of Metal-Organic Frameworks for Energy-Efficient Adsorption-Based Refrigeration Systems

Abstract:

Energy used for refrigeration, air conditioning and heat pump processes accounted for 17% of the global electricity consumption in 2015 and is expected to further increase. Adsorption-driven refrigeration has emerged as a promising and more energy-efficient alternative to traditional vapor-compression systems, since it can utilize lower-grade thermal energy such as solar energy and industrial waste heat. The choice of working fluid/adsorbent working pair is critical to the performance of an adsorption refrigeration system. Metal-organic frameworks (MOFs), due to their large surface areas and tunable chemical environments, have the potential to be excellent adsorbents for a variety of working fluids. Here, I will present my recent work on computational modeling and optimization of MOF-based adsorption refrigeration systems. In the first part, we tried to establish a structure-property relationship between the MOFs' structures and their working capacities, using ethanol as the working fluid. We focused on a set of zirconium-based MOFs with the same building blocks but different topologies and pore sizes. Using grand canonical Monte Carlo (GCMC) simulations, we were able to predict the working capacity of each MOF/ethanol working pair by simulating the ethanol adsorption isotherms in those MOFs. Our results suggest that working conditions with lower temperatures (such as ice-making) favor MOFs with smaller pores, and vice versa. Semi-quantitative relationships between the pore sizes and the working capacities were established and verified by experiments. In a follow-up study with propane and isobutane as working fluid (which have been recommended by EPA as freon substitutes), we were able to further validate our structure-property relationships. The working capacity of the NU-1003/isobutane working pair, one of the best pairs we discovered in this study, is 3 times larger than the MIL-101/isobutane pair which was previously the best one reported in the literature. We hope that our results can guide the design of materials with superior performance for adsorption-based refrigeration, which is a promising solution to reduce global energy cost and carbon emissions in the future.

597569 Advanced Monte Carlo Techniques for Simulating Xylene Isomer Separations in One-Dimensional Metal-Organic Frameworks

Abstract:

p-Xylene is one of the most important chemicals in the world. However, separation of p-xylene from the other xylene isomers is difficult because the isomers have similar physical properties, such as boiling points.[1] Previous studies have shown that porous materials such as zeolites and metal-organic frameworks (MOFs) can be useful for xylene separations,[2–4] and there are several proposed mechanisms for adsorption separation of xylenes. Torres-Knoop et al.[3,5] proposed the "commensurate stacking" mechanism in 1D channels, in which para-selectivity arises because p-xylene can use the space in the framework more effectively than the other isomers.

Here we present a computational screening study of the p-xylene selectivities using grand canonical Monte Carlo (GCMC) simulations of 1D channel structures from the ToBaCCo MOF database by Colon et al.[6] To overcome the difficulties in inserting xylene molecules in tight-fitting pores, we used the Energy-Biasing Monte Carlo (EBMC) scheme[7] to place the center of the aromatic ring and combined EBMC with orientational biasing of the rigid xylene molecules. We found that this combined biasing move is highly effective when dealing with insertions at saturation loading, when the packing is very tight, and the molecules are localized at specific locations.

Using this method, we were able to identify structures within the ToBaCCo database with high para-selectivity. Some of them present different packing patterns than what Torres-Knoop et al. reported,[3,5] and we conclude that face-to-face stacking is not the only way to achieve high para-selectivity. We observe mechanisms including parallel slit interaction, perpendicular stacking and steric hindrance.

We also propose a more coarse-grained model that can be used to significantly speed up the screening process. In the coarse-grained model, the aromatic ring is represented by a single "large" Lennard-Jones site with 2 attached methyl sites, creating a three-site model for the xylene isomers. We show that this simple model can mimic the behavior of the more detailed model[4] in GCMC simulations of adsorption, in particular for the selectivity and molecular packing in the 1D channels.

References:

(1) Sholl, D. S.; Lively, R. P. Seven Chemical Separations to Change the World. *Nature News* **2016**, *532* (7600), 435. https://doi.org/10.1038/532435a.

(2) Gee, J. A.; Zhang, K.; Bhattacharyya, S.; Bentley, J.; Rungta, M.; Abichandani, J. S.; Sholl, D. S.; Nair, S. Computational Identification and Experimental Evaluation of Metal–Organic Frameworks for Xylene Enrichment. *J. Phys. Chem. C* **2016**, *120* (22), 12075–12082. https://doi.org/10.1021/acs.jpcc.6b03349.

(3) Torres-Knoop, A.; Krishna, R.; Dubbeldam, D. Separating Xylene Isomers by Commensurate Stacking of P-Xylene within Channels of MAF-X8. *Angewandte Chemie International Edition* **2014**,

53 (30), 7774–7778. https://doi.org/10.1002/anie.201402894.

(4) Castillo, J. M.; Vlugt, T. J. H.; Calero, S. Molecular Simulation Study on the Separation of Xylene Isomers in MIL-47 Metal–Organic Frameworks. *J. Phys. Chem. C* **2009**, *113* (49), 20869–20874. https://doi.org/10.1021/jp908247w.

(5) Torres-Knoop, A.; Balestra, S. R. G.; Krishna, R.; Calero, S.; Dubbeldam, D. Entropic Separations of Mixtures of Aromatics by Selective Face-to-Face Molecular Stacking in One-Dimensional Channels of Metal–Organic Frameworks and Zeolites. *ChemPhysChem* **2015**, *16* (3), 532–535. https://doi.org/10.1002/cphc.201402819.

(6) Colón, Y. J.; Gómez-Gualdrón, D. A.; Snurr, R. Q. Topologically Guided, Automated Construction of Metal–Organic Frameworks and Their Evaluation for Energy-Related Applications. *Crystal Growth & Design* **2017**, *17* (11), 5801–5810. https://doi.org/10.1021/acs.cgd.7b00848.

(7) Snurr, R. Q.; Bell, A. T.; Theodorou, D. N. Prediction of Adsorption of Aromatic Hydrocarbons in Silicalite from Grand Canonical Monte Carlo Simulations with Biased Insertions. *J. Phys. Chem.* **1993**, *97* (51), 13742–13752. https://doi.org/10.1021/j100153a051.

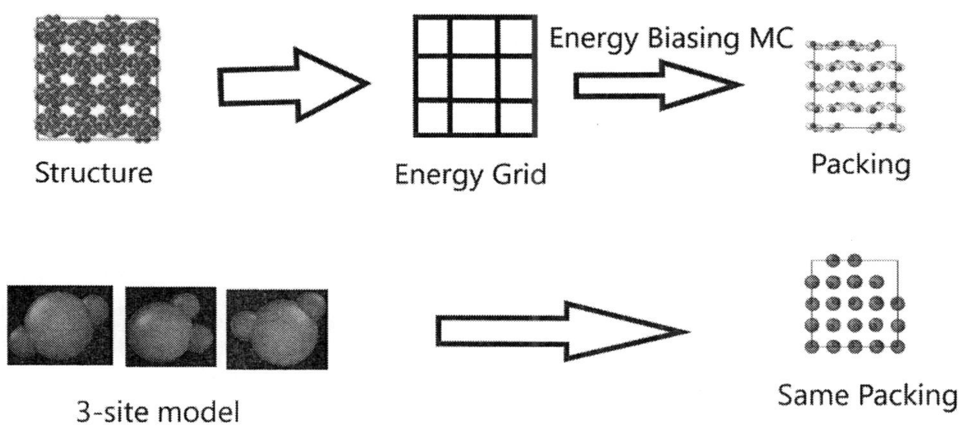

609363 Effectiveness of Free Energy and Other Crystal Descriptors to Assess MOF Synthesizability: To What Extent Is Thermodynamics Enough?

Abstract:

Metal-organic frameworks (MOFs) have garnered substantial attention for use in diverse applications, due mainly to their highly tunable pore structure and chemistry, which arises from the modular nature of their construction. However, this modularity also results in a combinatorial explosion of possible MOFs compositions, making comprehensive MOF screening challenging. High throughput computational screening has accelerated property prediction of computational MOF prototypes. However, which computational MOF prototypes are synthetically feasible (or likely) is a question that remains unanswered. One question still under debate is to what extent synthetic likelihood is determined by thermodynamic or kinetic factors, respectively. Previously, we demonstrated our capability of performing high throughput MOF free energy calculations by calculating the free energies in a large, computational MOF database [1]. This capability allows us to investigate whether thermodynamic factors correlate with the apparent synthesizability of prototypes in a MOF database. Apparent synthesizeability was assessed by finding which computational prototypes were also found in experimental repositories, e.g. the Cambridge database.

We begin with a detailed discussion regarding the differences between free energies calculated using thermodynamic integration and the quasiharmonic approximation, as a function of temperature, and highlighting the limitations of the quasiharmonic approximation. Next, we use our calculated free energies to first define a region of "global stability" where all previously synthesized MOFs within our database seem to cluster. This allows us to create a filter where MOFs outside of this region are deemed synthetically unlikely. However, we found this filter to be insufficient when multiple MOF isomorphs all exist within the region of global stability. We find that, in 80% of the cases, the previously synthesized member of an isomorphic series of MOFs has the lowest calculated free energy, revealing that, in the majority of cases, thermodynamics will correctly predict synthetically accessible isomorphs. Furthermore, all previously synthesized isomorphs have a free energy within 3% of that of the calculated lowest free energy isomorph. This makes free energy relative to the lowest free energy isomorph a reasonable second filter to eliminate synthetically inaccessible MOF prototypes within an isomorphic series. Therefore, although kinetics are bound to play a role in certain cases, the application of a thermodynamic filter can be effective in preventing futile synthesis attempts on MOF prototypes that are otherwise promising, but that can be classified as synthetically inaccessible. Finally, we discuss how the effectiveness of free energy "filters" change with temperature, and how filters based on other, easier-to-calculate MOF properties perform relative to free energy.

[1] Anderson, R.; Gómez-Gualdrón, D. A. Large-Scale Free Energy Calculations on a Computational MOF Database : Toward Synthetic Likelihood Predictions. *ChemRxiv* **2020**, 1–16.

603380 An Atomistic Lattice Dynamics Approach for Free Energy Calculations within Crystal Structure Prediction Studies

Abstract:

A plethora of organic molecules exhibit polymorphism, which refers to the ability of chemical compounds to pack into different crystalline motifs. This phenomenon is of special importance both to industry and academia since physical and chemical properties, such as solubility, bioavailability and mechanical strength may vary tremendously between polymorphs. From a thermodynamic standpoint, polymorphs can be identified as minima on the free energy (FE) landscape, with the most stable form corresponding to the global minimum and other forms corresponding to local minima (metastable structures). This thermodynamic understanding has motivated the development of crystal structure prediction (CSP) tools that are designed to determine all polymorphs for a given compound with the correct order of stability based on minimal information, such as the chemical connectivity diagram[1].

Recent advances in CSP were highlighted in the last blind test organised by Cambridge Crystallographic Data Centre[2]. It is worth noting that only 7 out of the 25 groups participating in the last test have incorporated FE calculations within their workflow, while the remaining groups used only lattice energy in their predictions, thus neglecting temperature and vibrational effects. Lattice dynamics (LD) theory was deployed successfully for the evaluation of vibrational free energies, utilizing either dispersion-corrected periodic density functional theory[3,4] (DFT-d) or force field methods based on distributed multipoles expansion[5,6,7] (DMA). DFT-d can provide very accurate results at a high computational cost, whereas the DMA-based approach provides a good trade-off between accuracy and efficiency but cannot account for internal modes arising from intramolecular vibrations. A limitation of both methods is that they rely on the construction of supercells, which increases computational demands and results in some ambiguity in the generation of dispersion curves.

In this work, we present a recently-developed methodology[8] for performing atomistic lattice dynamics calculations, which is suitable for conducting FE calculations for a large number (>100) of structures generated in CSP studies. This methodology is based on LD theory within the harmonic approximation, with intramolecular interactions quantified by isolated quantum mechanical calculations[9]. The intermolecular interactions are partitioned into electrostatics and repulsion/dispersion interactions and are modelled using *ab initio* distributed multipoles and a semi empirical potential, respectively. To illustrate the approach, we apply a multistage CSP methodology to tetracyanoethylene (C_6N_4), a small planar molecule. The global search stage is performed using the CrystalPredictorII[10,11] for the generation of candidate structures. The lowest energy structures, within 20kJ/mol from the global minimum in lattice energy, are minimised atomistically using CrystalOptimizer[12] yielding 837 unique structures after clustering. We evaluate the vibrational free energy of the 200 most promising structures, using our in-house LD algorithm[8], for the determination of polymorphic stability, entropy, heat capacity and thermal energy in a temperature range between 0 and 400K.

References:

(1) Pantelides, C. C.; Adjiman, C. S.; Kazantsev, A. V. General Computational Algorithms for Ab Initio Crystal Structure Prediction for Organic Molecules. In *Prediction and Calculation of Crystal Structures Methods and Applications. Topics in Current Chemistry*; Atahan-Evrenk, S.; Aspuru-Guzik, A., Eds.; Springer, 2014; Vol. 345, pp. 25–58.

(2) Reilly, A. M.; Cooper, R. I.; Adjiman, C. S.; Bhattacharya, S.; Boese, A. D.; Brandenburg, J. G.; Bygrave, P. J.; Bylsma, R.; Campbell, J. E.; Car, R.; *et al.* Report on the Sixth Blind Test of Organic Crystal Structure Prediction Methods. *Acta Crystallogr. Sect. B Struct. Sci. Cryst. Eng. Mater.* **2016**, *72*, 439–459.

(3) Hoja, J.; Ko, H.-Y.; Neumann, M. A.; Car, R.; DiStasio, R. A.; Tkatchenko, A. Reliable and Practical Computational Prediction of Molecular Crystal Polymorphs. **2018**, 1–9.

(4) Brandenburg, J. G.; Grimme, S. Organic Crystal Polymorphism: A Benchmark for Dispersion-Corrected Mean-Field Electronic Structure Methods. *Acta Crystallogr. Sect. B Struct. Sci. Cryst. Eng. Mater.* **2016**, *72*, 502–513.

(5) Nyman, J.; Day, G. M. Static and Lattice Vibrational Energy Differences between Polymorphs. *CrystEngComm* **2015**, *17*, 5154–5165.

(6) Price, S. L.; Leslie, M.; Welch, G. W. A.; Habgood, M.; Price, L. S.; Karamertzanis, P. G.; Day, G. M. Modelling Organic Crystal Structures Using Distributed Multipole and Polarizability-Based Model Intermolecular Potentials. *Phys. Chem. Chem. Phys.* **2010**, *12*, 8466.

(7) Stone, A. J. Distributed Multipole Analysis, or How to Describe a Molecular Charge Distribution. *Chem. Phys. Lett.* **1981**, *83*, 233–239.

(8) Vasileiadis, M. Calculation of the Free Energy of Crystalline Solids, Imperial College London, 2013.

(9) Frisch, M. J.; Trucks, G. W.; Schlegel, H. B.; Scuseria, G. E.; Robb, M. A.; Cheeseman, J. R.; Scalmani, G.; Barone, V.; Mennucci, B.; Petersson, G. A.; Nakatsuji, H.; Caricato, M.; Li, X.; Hratchian, H. P.; Izmaylov, A. F.; Bloino, J.; Zheng, G.; Sonnenb, D. J. al. Gaussian 09, 2009.

(10) Habgood, M.; Sugden, I. J.; Kazantsev, A. V.; Adjiman, C. S.; Pantelides, C. C. Efficient Handling of Molecular Flexibility in Ab Initio Generation of Crystal Structures. *J. Chem. Theory Comput.* **2015**, *11*, 1957–1969.

(11) Sugden, I.; Adjiman, C. S.; Pantelides, C. C. Accurate and Efficient Representation of Intramolecular Energy in Ab Initio Generation of Crystal Structures. I. Adaptive Local Approximate Models. *Acta Crystallogr. Sect. B Struct. Sci. Cryst. Eng. Mater.* **2016**, *72*, 864–874.

(12) Kazantsev, A. V.; Karamertzanis, P. G.; Adjiman, C. S.; Pantelides, C. C. Efficient Handling of Molecular Flexibility in Lattice Energy Minimization of Organic Crystals. *J. Chem. Theory Comput.* **2011**, *7*, 1998–2016.

606393 Accelerated Saddle Point Refinement with Sella

- **Download/View Extended Abstract**

Abstract:

Identification and refinement of first order saddle point (FOSP) structures on the potential energy surface (PES) of chemical systems is a computational bottleneck in the characterization of reaction pathways. For small systems, it may be feasible to evaluate the full Hessian, which determines both the most probable direction of the reaction coordinate and the optimal step size and direction for refinement to the FOSP. However, evaluation of the full Hessian quickly becomes prohibitively expensive as the system size grows. Thus, many FOSP refinement methods designed for use with larger systems (such as those relevant to heterogeneous catalysis) do not evaluate the Hessian in full. Instead, these approaches calculate the lowest-curvature mode of the Hessian using an iterative diagonalization procedure. These procedures approximate Hessian-vector products by performing finite-difference on the gradient vector. Crucially, the full Hessian is no longer available in this approach for the determination of optimal geometry steps during FOSP refinement, which are instead determined using an approximate Hessian. This drastically increases the number of geometry steps required to converge to a FOSP, and may result in a failure to converge all-together.

In a recent publication[1], we describe a FOSP refinement procedure that accelerates convergence to a saddle point using iterative Hessian diagonalization. The Hessian-vector products that have been evaluated during diagonalization are used to construct a highly accurate approximate Hessian. In the full-diagonalization limit, the approximate Hessian constructed in this way becomes exact. The total number of diagonalization steps performed is controlled by a single convergence parameter, γ. Tighter convergence yields a more accurate Hessian – thereby reducing the number of refinement steps needed to locate the FOSP – at the cost of additional diagonalization iterations. The cost of diagonalization iterations can be balanced against the cost of additional geometry refinement steps by carefully tuning the value of γ.

Our method has been implemented in an open source software package, Sella[2]. Sella is written in Python and is compatible with all major operating systems and computer architectures. Sella uses the ASE[3] library to interface with a wide range of software packages for PES evaluations, including electronic structure theory packages and classical force-field packages. In particular, we have used the ASE interface to LAMMPS[4] for several test systems.

We measure the performance of our approach as implemented in Sella on two FOSP refinement benchmarks from optbench.org[5]. On one benchmark, we see an average reduction of over 50% in the number of gradient evaluations (consisting of diagonalization iterations plus geometry refinement steps) required to converge to a FOSP. On a second benchmark, we see an average reduction in the number of gradient evaluations of over 25%. Additionally, our implementation of iterative diagonalization requires an average of 30% fewer iterations to converge to the leftmost eigenvector of the Hessian compared to the best performing codes listed on optbench.org. When an approximate Hessian is available, our Jacobi-Davidson-based diagonalization routine requires on average 40% fewer iterations to converge than non-preconditioned Lanczos. These performance improvements are a direct result of the increased accuracy of the approximate Hessian matrix that our method enables.

[1] Hermes, E.D., Sargsyan, K., Najm, H.N., Zádor, J. *J. Chem. Theory Comput.* 15, 6536-6549

(2019).

[2] Hermes, E.D. *Sella,* 2019, DOI: 10.5281/zenodo.3379094.

[3] Larsen, A.H., Mortensen, J.J., Blomqvist, J., Castelli, I.E., Christensen, R., Dułak, M., Friis, J., Groves, M.N., Hammer, B., Hargus, C., Hermes, E.D., Jennings, P.C., Jensen, P.B., Kermode, J.B., Kitchin, J.R., Kolsbjerg, E.L., Kubal, J., Kaasbjerg, K., Lysgaard, S., Maronsson, J.B., Maxson, T., Olsen, T., Pastewka, L., Peterson, Al., Rostgaard, C., Schiøtz, J., Schütt, O., Strange, M., Thygesen, K.S., Vegge, T., Vilhelmsen, L., Walter, M., Zeng, Z., Jacobsen, K.W. *J. Phys.: Condens. Matter* 29, 273002 (2017).

[4] Plimpton, S. *J. Comput. Phys.* 117, 1 (1995).

[5] Chill, S.T., Stevenson, J., Ruehle, V., Shang, C., Xiao, P., Farrell, J.D. and Wales, D.J. *J. Chem. Theory Comput.* 10, 5476 (2014).

This work was supported by the U.S. Department of Energy, Office of Science, Basic Energy Sciences, Chemical Sciences, Geosciences and Biosciences Division, as part of the Computational Chemistry Sciences Program (Award Number: 0000232253).

Sandia National Laboratories is a multimission laboratory managed and operated by National Technology and Engineering Solutions of Sandia, LLC., a wholly owned subsidiary of Honeywell International, Inc., for the U.S. Department of Energy's National Nuclear Security Administration under contract DE-NA0003525. The views expressed in this article do not necessarily represent the views of the U.S. Department of Energy or the United States Government.

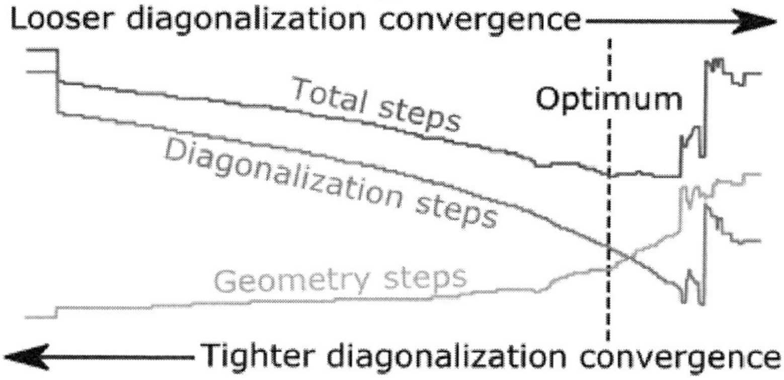

597834 Revisiting Electrochemical CO_2 Reduction on Copper Via Embedded Correlated Wavefunction Theory

Abstract:

Electrochemical CO_2 reduction to useful chemicals could contribute in a circular economy to reducing greenhouse gas emissions. Copper remains the only metal catalyst for CO_2 reduction to produce hydrocarbons but it requires a high overpotential for reasonable rates and faradaic efficiency. In addition, copper exhibits poor selectivity to useful products. Quantum mechanics simulations can aid in understanding the reaction mechanisms of CO_2 reduction on copper, thus providing insights for rational design of more efficient electrocatalysts. All theoretical efforts in this area to date have been within the framework of approximate density functional theory (DFT), which suffers from electron self-interaction error. DFT predicts that the key intermediate, CO, adsorbs in a hollow site while atop site adsorption has been observed experimentally on copper, suggesting that DFT-predicted reaction mechanisms and energetics could be problematic. Instead, embedded correlated wavefunction (ECW) theory, which provides a local correction to exchange-correlation error inherent in DFT, successfully reproduces the observed CO atop site adsorption. We are the first to apply ECW theory to study electrochemical CO_2 reduction on copper. Within ECW, the extended surface is described by an embedding potential derived from density functional embedding theory, while the catalytically active center is treated with CW theory in the presence of the derived embedding potential. We will present ECW-predicted reaction mechanisms of CO_2 reduction on copper, including the critical CO reduction and C-C bond formation steps. We will compare rate-limiting steps predicted by DFT and ECW theory, as well as structural and mechanistic behavior predicted by these two methods.

611957 Explicit Solvent Hybrid QM Simulations Embedded in the Quantum-Based Polarizable Reactive Force Field

Abstract:

Multiscale reactive simulations of materials and energy systems at the accuracy level of ab-initio methods are required to make significant breakthroughs in the design and discovery of novel complex materials. Hybrid quantum mechanics/molecular mechanics (QM/MM) methods have often been used to enable QM accuracy for the key reactive part of a system while using a force field for the dynamics of more distant atoms. Generally the electrostatics and polarizations across the boundary of the QM and MM subsystems is poorly described, limiting the accuracy. Recently, we developed a new generic reactive force field, RexPoN, capable of describing complex reaction dynamics including polarization at nearly the accuracy of QM for systems with 10,000 to 1 million atoms, which we have validated with several applications. Here, we introduce a new hybrid QM/MM framework, RexPoN embedded QM, or ReQM, for practical multi-scale simulations of systems containing millions of atoms, but for which the long range interactions in the polarizable Reactive field (PRD) match accurately those of QM. ReQM is suitable for applications such reactive simulations as electrocatalysis, that require accurate full explicit description of the solvent . We validate the accuracy of ReQM by computing the vibrational density of states of reaction intermediates during CO_2 reduction reactions and comparing against operando electrocatalysis and full solvent QM calculations.

600197 Predicting the Structure of Lanthanide-Ligand Complexes in Solution with Ab Initio Molecular Dynamics

Abstract:

Rare earth elements, or lanthanides, are critical in many clean energy and defense technologies. Lanthanide-ligand complex structures are relevant to their separation by solvent extraction. We recently published lanthanide norm-conserving pseudopotentials and basis sets optimized for the generalized gradient approximation and exchange correlation functionals for density functional theory calculations of large systems, which can replicate lanthanide reaction energies and molecular geometries. These allowed us to perform ab initio molecular dynamics simulations of lanthanide aqua ions with explicit solvent molecules. Results agree with experiment showing coordination numbers of nine and eight for the early and late lanthanides respectively, and lanthanide-oxygen radial pair distributions are within 0.05 Å of experimental values. The first coordination shell shows a dynamic equilibrium and not a rigid geometry. Preliminary results also show that we are replicating lanthanide-ligand structures. Our simulations predict the known europium-EDTA coordination structure in basic conditions, with both nitrogen atoms and four EDTA oxygen atoms (one per carboxylate) coordinated to Eu, and an additional three coordinated water molecules. More importantly, the predicted Eu-O and Eu-N distances are within ~0.05 Å of X-ray absorption experiments. Currently, we are exploring the binding behavior of phytosiderophore ligands; preliminary computational and experimental results show lanthanide binding.Our results strongly suggest that we can obtain meaningful structures of lanthanide-ligand complexes in the condensed phase. We report a computational capability that can accurately model lanthanide-ligand complex structures and reactions in the solution phase.

606186 Development of New Coarse-Grained Water Model Using a Hybrid Coarse-Graining Approach

Abstract:

Water is one of the most common solvent used for industrial purposes. Development of an accurate and efficient water molecule force field for use in the molecular dynamics (MD) simulations is a challenging task. There is a number of different models for bulk water that are available for use in MD simulations. However, none of the available models are capable of satisfying the following three requirements at once: accuracy, efficiency, and simplicity. The objective of this work is to develop a computationally efficient force field for a water model from *ab initio* quantum chemistry data. This new model would accurately predict structure, dynamic, thermodynamic, and transport properties. To perform this task, multi-scale coarse-graining approach (MSCG) has been applied to *ab initio* quantum chemistry data. In MSCG atoms are grouped into beads, thus, decreasing the number of degrees of freedom. For an accurate structure prediction a new hybrid scheme has been applied combining MSCG method with an iterative Boltzmann inversion (IBI) coarse-graining method. The model developed by MSCG+IBI can be used to predict structure and thermal properties. However, it can't predict the dynamics and transport properties of the system. To remedy this, a probability distribution function coarse-graining (PDF-CG) method has been used, which recovers the dynamics and transport properties. By combining MSCG+IBI and PDF-CG methods, a highly accurate and computationally efficient water model has been developed. This model may be used to perform coarse-grained molecular dynamics simulations of aqueous solutions of various liquids of interest.

608558 Let's Not Use Histograms: Bayesian Inference of Potentials of Mean Force with Umbrella Sampling and Multistate Reweighting

Abstract:

Potentials of mean force (PMFs)—free energies along a selected set of collective variables—are ubiquitous in molecular simulation, and of significant value in understanding and engineering molecular behaviors. PMFs are most commonly estimated using variants of histogramming techniques, but such approaches obscures two important facets of the distribution. First, the empirical observations along the collective variable are defined by an ensemble of discrete observations and the coarsening of these observations into a histogram bins incurs unnecessary loss of information. Second, the potential of mean force is itself almost always a continuous function, and its representation by a histogram introduces inherent approximations due to the discretization. In this study, we relate the observed discrete observations to the underlying continuous probability distribution over the collective variables and derive histogram-free estimation techniques for the potential of mean force.

We reformulate PMF estimation to minimization of a Kullback-Leibler divergence between a continuous trial function and the discrete empirical distribution and show this is equivalent to likelihood maximization of a trial function and the sampled data. We then present a fully Bayesian treatment of this formalism to enable the incorporation of powerful Bayesian tools such as the inclusion of regularizing priors, uncertainty quantification, and model selection techniques. We demonstrate our new formalism in the analysis of umbrella sampling simulations for the χ torsion of a valine sidechain in the L99A mutant of T4 lysozyme with benzene bound in the cavity.

594434 Entropy Calculation from Molecular Dynamics Simulation

Abstract:

In classical molecular dynamics (MD) simulation, certain thermodynamic properties, including temperature, pressure and internal energy can be computed in a routine matter during the course of the simulation. Entropy and properties such as the Helmholtz free energy or the chemical potential, which depend on entropy, require significantly more sophisticated approaches to quantitatively evaluate. The ultimate objective of this work is to develop a software module for LAMMPS or other classical MD simulator from which the entropy can be calculated with an effort on par to that of other thermodynamic properties. The approach to this problem is to develop a theoretical expression for excess entropy as a universal functional of the radial distribution function (RDF), which can be routinely calculated from an MD simulation. The functional builds upon the modified Kirkwood entropy in the fluid phase and a local approximation to the harmonic crystal in the solid phase. Five materials with vastly different interaction potentials are simulated across solid and fluid phases from 1 K to 10^7 K. The standard for entropy, by which the validity of the entropy functional is judged, is obtained from conventional thermodynamic integration over this temperature range. Comparison of the functional used to reproduce the entropy standard for the five materials provides insight into the requirements for universality of the functional. Solutions to address numerical challenges to the robust and accessible implementation of the entropy calculation are shared.

597813 Molecular Design Blueprints: Catalysts and Principles from New Simulation and Machine Learning Tools

Abstract:

Many highly selective catalysts have been discovered that are defined by well-isolated sites with tailored metal-organic bonding. The rational design of such sites in de novo transition metal complexes, metal-organic frameworks (MOFs), or single atom catalysts, however, remains challenging. First-principles (i.e., with density functional theory, or DFT) high-throughput screening is a promising approach but is hampered by high computational cost, particularly in the brute force screening of large numbers of materials. In this talk, I will outline our efforts over the past few years to accelerate the design of single-site catalysts. I will describe our software and machine learning (ML) models that both simplify and accelerate the screening of new materials. These tools have enabled us to uncover new design rules and exceptions in both molecular and periodic materials (e.g., MOFs). We have paired ML models with multiobjective optimization strategies, robust uncertainty quantification, and highly parallel accelerated computing to discover optimal materials in weeks instead of decades. Time permitting, I will describe our recent efforts in autonomous computational chemistry by developing ML models as decision engines capable of predicting when calculations will fail and when methods beyond DFT are needed.

600080 Machine Learning Quantum Tunneling in the Kinetics of Chemical Reactions

Abstract:

The computation of reaction rate constants has been tackled using a multitude of theories ranging from classical transition state theory[1] to the fully quantum flux-flux correlation function approach.[2] The main cost of these methods is related to the need to explore potential energy surfaces, PES, either at the level of geometry optimizations, minimum energy path searches or to describe dynamics in time. In this context, machine learning algorithms which accelerate the search for reactive pathways or the prediction of rate constants are of great interest.[3,4] We have constructed a database of over 130000 reaction rate constants[5] obtained via quantum calculations of the reaction rate constants for particles crossing one dimensional single and double symmetric and asymmetric barriers. A deep neural network, DNN, was trained on the database with the ADAM optimizer and used to predict reaction rate constants. The most significant input features were identified using the Pearson correlation coefficient, and a grid search analysis was carried out for a coarse optimization of the DNN model hyperparameters. The network was trained for hundreds of epochs. The predicted results from the test set are in good agreement with the exact values. This approach shows promise for the application of machine learning to predict reaction rate constants.

1. Pechukas, P. Transition State Theory. *Annu. Rev. Phys. Chem.* **32**, 159–177 (1981).
2. Tromp, J. W. & Miller, W. H. The reactive flux correlation function for collinear reactions H + H2, Cl + HCl and F + H2. *Faraday Discuss. Chem. Soc.* **84**, 441–453 (1987).
3. Amabilino, S. *et al.* Training neural nets to learn reactive potential energy surfaces using interactive quantum chemistry in Virtual Reality. *J. Phys. Chem. A* **123**, 4486–4499 (2019).
4. Ulissi, Z. W. *et al.* Machine-learning methods enable exhaustive searches for active Bimetallic facets and reveal active site motifs for CO2 reduction. *ACS Catal.* **7**, 6600–6608 (2017).
5. Valleau, S. Machine learning quantum tunneling in the kinetics of chemical reactions. *In preparation.* (2020).

611092 Improved Catalyst Design and Performance through Application of Machine Learning Algorithms

- Download/View Extended Abstract

Abstract:

Conversion of biomass feedstocks to hydrocarbon fuels is often fraught with catalyst deactivation due to deposition of coke on catalyst surface or deactivation due to inorganic components of the biomass. Acidic zeolite catalyst get easily deactivated because of the basic oxides in the biomass feedstock. The high aromatic content and macromolecules in the biomass pyrolysis oils also tend to promote carbon formation on the active sites, which eventually lead to their deactivation. We developed a nickel supported on mixed oxide (Ni/RM) catalyst that is more robust than the commercial nickel on silica-alumina catalyst. This new catalyst is regenerable and can be used effectively for hydrodeoxygenation of biomass pyrolysis oils to long chain hydrocarbon fuels. The new catalyst is effective in opening both furan and benzene rings and thus producing hydrocarbon fuels with linear and branched carbon chain length between C6 to C15. The catalyst was not deactivated by inorganic components in the biomass, however, the catalyst activity decreased with deposition of coke on the active sites and required frequent catalyst regeneration through combustion and reduction to restore activity. We developed a machine learning (ML) algorithm which enabled us to reformulate the catalyst and reduce the carbon deposition on the active site. The *signature* of each nickel supported mixed oxide catalyst (Ni/RM)) was defined as the weight percent of *eleven elements* (Al, Ca, Fe, K, Mg, Na, Ni, P, S, Si, Ti). These together with the *reaction temperature formed the feature vector* (input to ML). The output of the ML was coke formation, which was also a performance measure for the designed catalyst. The network relating the ML inputs to the output was either a simple Neural Network or a Convolutional Neural Network (CNN). Using this ML model as a basis, we inverted the model (inverse modeling) to obtain a new Ni/RM catalyst, which was then validated through experiments. The machine learning approached reduced catalyst development time and improved the robustness of the catalyst. Thus, the carbon deposition rate was reduced and catalyst regeneration frequency was also reduced.

603620 Improving the Reusability of Machine Learning Models for Materials

Abstract:

Machine learning models have proven useful in a variety of chemical and materials applications. However, one drawback of these models is that they are often geared to a particular application, and thus a new model must be created for a new application. This is particularly noticeable in applications that feature guest-host interactions, broadly defined, as there is a huge combinatorial challenge in considering many possible guests and many possible hosts. For example, this combinatorial challenge applies to catalysis, adsorption, doping, and intercalation. In this work, we develop and apply strategies to improve the generality and reusability of machine learning models.

We have developed a machine-learning architecture that takes advantage of inherent aspects of chemical systems, and is particularly geared towards those that feature guest-host interactions. This framework partially decouples the guests from the hosts by introducing intermediate variables, which are a set of implicit host properties that control how the host interacts with all guests. These intermediate variables are learned during the fitting process. We create separate sub-models for different host elements for predicting these intermediate variables, and separate sub-models for how different guests respond to these host variables. The sub-models are all fit simultaneously. Our framework takes advantage of the fact that elements are discrete entities, and greatly simplifies the huge combinatorial challenge of considering many possible guests and many possible hosts. We apply this framework to multiple datasets and show how it can be used to improve the efficiency and accessibility of initial screening in a variety of applications.

610067 The Open Catalyst Project Dataset

Abstract:

The Open Catalyst Project aims to develop new ML methods and models to accelerate the catalyst simulation process for renewable energy technologies and improve our ability to predict activity/selectivity across catalyst composition. To achieve that in the short term we need participation from the ML community in solving key challenges in catalysis. One path to interaction is the development of grand challenge datasets that are representative of common challenges in catalysis, large enough to excite the ML community, and large enough to take advantage of and encourage advances in deep learning models. Similar datasets have had a large impact in small molecule drug discovery, organic photovoltaics, and inorganic crystal structure prediction. We present the first open dataset from this effort on thermochemical intermediates across stable multi-metallic and p-block doped surfaces. This dataset includes full-accuracy DFT calculations across 53 elements and their binary/ternary materials, various low-index facets. Adsorbates span 56 common reaction intermediates with relevance to carbon, oxygen, and nitrogen thermal and electrochemical reactions. Off-equilibrium structures are also generated and included to aid in machine learning force field design and fitting. Collectively, this dataset represents the largest systematic dataset that bridges organic and inorganic chemistry and will enable a new generation of catalyst structure/property relationships. Fixed train/test splits that represent common chemical challenges and an open challenge website will be discussed to encourage competition and buy-in from the ML community.

607970 Active-Learning Framework for the Discovery of Novel Materials for Catalysis

Abstract:

The discovery of abundant and affordable materials for sustainable energy applications is a significant challenge of this decade. The space of possible material compositions and crystal structures is so vast that brute-force screening approaches are unfeasible. However, density functional theory (DFT) combined with machine-learning (ML) techniques have a huge potential to speed up the search and hold a great promise for the discovery of novel materials.

Here, we present a ML-based active-learning framework to search for stable and meta-stable inorganic materials. The workflow includes the generation of candidate structures, which is followed by automated DFT job submission, ML-training, and acquisition. When using a search space of experimentally observed crystal structures, this approach was recently applied to the discovery of new stable polymorphs of IrO_x, predicted to have an enhanced stability and catalytic activity for the oxygen evolution reaction (OER). In this talk, we extend the search to hypothetical crystal structures that are generated from a bottom-up prototype enumeration scheme, based on spacegroup and Wyckoff positions. This is applied to the search for cheaper and more abundant materials. Last, we envisioned how the screening approach can be extended to surface and chemisorption properties, on order to extend the search to catalytic applications.

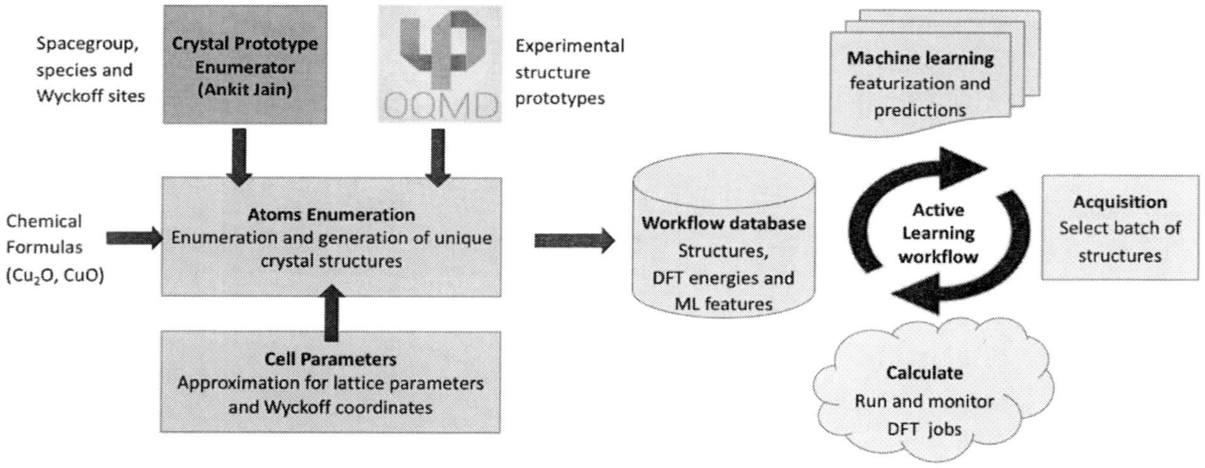

606321 Identifying Interaction Trends between Single Metal Atoms and Modified MgO(100) Supports with Statistical Learning

Abstract:

In this work, we apply statistical learning (SL) to identify physical descriptors for predicting electronic metal-support interactions (EMSI) in catalysis. Charge transfer between metal atoms and their underlying oxide support influences both the binding strength of the metal to the support and the binding strength of reaction intermediates to the metal. Tuning charge transfer by introducing surface-modifying dopants or co-adsorbates is an important strategy for controlling catalytic behavior. We use SL to build models for predicting metal atom binding energy to the oxide surface as a function of readily available physical descriptors containing properties of the metal, the support, and the surface-modifiers. Models are trained against DFT datasets consisting of metal adsorption energies on modified MgO(100) surfaces featuring various aliovalent dopants (e.g., Al, B, Li, and Na) and adsorbates (e.g., F, H, OH, and NO_2). These modifications generate both electron-rich and electron-poor MgO surfaces, which in turn enhance the extent of charge transfer between the metal and the support. We also test multiple SL procedures, including LASSO, Horseshoe prior, and Dirichlet-Laplace prior, and find that features derived by Dirichlet-Laplace prior construct the most accurate predictive models (i.e., achieving RMSE of ~0.2 eV for predicted binding energies). Moreover, we find that the features selected by SL using MgO training data generally are transferable to similar oxides, such as CaO(100), BaO(100), and ZnO(100), where data for these oxides are excluded from the feature selection process. This demonstrates the robust nature of features selected by the Dirichlet-Laplace prior procedure, both in terms of model accuracy and transferability to oxide surfaces beyond those included in the training set.

611372 How to Fix Some of Our Errors Via Data Science

Abstract:

Over the past two decades, multiscale modeling has advanced tremendously, and several algorithms currently exist. Yet, our ability to apply first principles modeling to process design is seriously limited due to multiple challenges. One of the challenges relates to error of first principles calculations, reaction mechanisms, active sites, etc. and its impact on catalyst discovery and reactor design. In this talk we will discuss sources of errors and new statistical learning methods that are transferable and explainable to quantify and correct some of the errors. We will discuss data quality and probabilistic graph models as a framework that integrates data, knowledge, and physical models to enable uncertainty quantification.

609630 Analysis of Binary Adsorption into a Surfactant Bilayer Using Gibbs Ensemble Monte Carlo Simulations and Ideal Adsorbed Solution Theory

Abstract:

Solute uptake into microstructured soft materials is a complex phenomenon with applications spanning the delivery of therapeutics to the release of personal care products. Previous attempts to simulate this process have successfully estimated the critical density of solute that could load into a surfactant system, but determining structural details at the molecular level has proved challenging. Using configurational-bias Monte Carlo simulations in the osmotic Gibbs ensemble, we have been able to determine the loading of unary n-nonane and 1-hexanol into a triethylene glycol mono-n-decyl ether (C10E3) surfactant bilayer, as well as the resulting structural effects caused by each solute. However, due to solute interactions, adsorption becomes more complex in multicomponent systems. Here, we apply a similar method to the binary adsorption of both n-nonane and 1-hexanol into the same surfactant bilayer system, allowing us to determine the partial molar properties of each solute and compare the favorability of molecular interactions within this bilayer system. We further investigate these solute interactions by comparing simulation to the behavior predicted by Ideal Adsorbed Solution Theory.

603868 Pushing the Frontier of Organic Optoelectronics through Computational Design of Open-Shell Donor-Acceptor Macromolecules

Abstract:

Low-bandgap optoelectronic materials find numerous applications in photovoltaics, sensors, magnetism, and photonics. Due to the limitless chemical space for carbon-based materials, in principle, one can design materials to meet any target functionality. Here, we highlight our success in designing small molecules and conjugated polymers for novel optoelectronic applications. We will share our work elucidating the role of diradical character in organic photovoltaics through careful selection of donor-acceptor (DA) moieties. We also show how control of the topology of DA macromolecule can lead to significantly high polyradical character and global aromaticity. We will conclude the presentation by discussing the realization of stable polymers with triplet ground state that display unique magnetic properties.

603411 Molecular-Level Insights into How Liquid Water Influences Thermodynamics at Water/Heterogeneous Catalyst Interfaces

Abstract:

Multiple societally important chemical reactions rely on catalytic processing in aqueous conditions, including biomass processing, electrochemical energy conversion, fertilizer production, and water purification. A goal of our work is to learn the molecular-level ways in which solvent molecules influence catalysis, both so that we can better understand catalyst fundamentals and so that we can garner insight needed to design new catalysts and optimize catalytic operating conditions. Experiments and simulations have uncovered a variety of ways in which water solvent influences catalytic phenomena. For example, it alters reaction intermediate and transition state energies, co-catalyzes certain reaction steps, and controls which catalytic pathways are followed. However, a comprehensive picture about how H_2O molecules influence catalytic behavior, including their influence on catalytic thermodynamics, remains unresolved. In this work, we use multiscale modeling involving density functional theory (DFT) and force-field molecular dynamics (ffMD) to examine how interfacial water influences the free energies of hydrophilic and hydrophilic species over hydrophobic Pt slabs, Pt electrodes, and supported Pt surfaces. We highlight how the polarity and polarizability of the solvent and the catalyst interface influence enthalpy and entropy of solvation and provide insight into how this influences catalytic phenomena.

611326 Computational Insights into Nanobiotechnology

Abstract:

Nanoscale interactions lie at the core of biological function and their understanding promises to bring major advances in different areas such as healthcare and the environment. It is also strongly believed that such an understanding will allow the design of new synthetic platforms able to reproduce the behavior of biological systems or interface with them in a seamless way. In this talk I will present our work on mimicking some of the properties of biological systems using nanoparticles, with an emphasis on the role that interactions with biological membranes can have. We have discovered that these nanoparticles can mimic several different functions performed by membrane proteins such as fusion proteins and transport proteins, opening new possibilities in delivering drugs, as well as serving as "artificial membrane proteins" themselves. It is important to highlight that our current molecular understanding of this system and the discoveries that came along were only possible due to careful molecular simulations that linked the observed macroscopic behavior to the microscopic world. To finalize I will give a perspective on exciting new avenues of research in this area.

605774 Model for Intrinsically Disordered Proteins with a Strong Dependence of Liquid-Liquid Phase Separation on Sequence

Abstract:

Phase separation of intrinsically disordered proteins is important for the formation of membraneless organelles, or biomolecular condensates, which play key roles in the regulation of biochemical processes within the cells. In this talk, we present the phase separation behavior of different sequences of a coarse-grained model for intrinsically disordered proteins and show that they exhibit a surprisingly rich phase behavior. Both the fraction of total hydrophobic parts and the distribution hydrophobic parts are investigated. We observed not only conventional liquid-liquid phase separation, but also reentrant phase behavior, in which the liquid phase density decreases at lower temperatures. For some sequences, we also observe formation of open phases consisting of aggregates, rather than a normal liquid. These aggregates had overall lower densities than the conventional liquid phases and complex geometries with large interconnected fibril-like or membrane-like clusters. Minor alterations in the ordering of a given set of amino acids may lead to large changes in the overall phase behavior, a fact of significant potential relevance for biology.

612426 Datamining Multicomponent Active Matter

Abstract:

Systems which convert energy into motion on the local scale exhibit a variety of emergent, steady-state behaviors. From a materials standpoint, this raises the possibility of a single 'active' material which has the ability to perform many, diverse, functions. However, leveraging non-equilibrium matter in new technologies necessitates an understanding of what, if any, quantity equilibrates at steady-state. We probe two fundamental sources of heterogeneity in multicomponent mixtures: distinct swim-speeds and distinct softness. For mixtures of fast and slow particles, we discover two regimes of behavior one with segregated domains of fast and slow species and one homogeneous. We show that a weighted average of constituent particle activities, which we term the net activity, defines a binodal for the motility-induced phase separation transition in active binary mixtures with distinct speeds and examine the critical point. Varying the softness of the active particles, we present an analytical approach which provides the density of the dense phase as well as the pressure. We also find that the surface tension, computed from equilibrium quantities, has a linear relationship with activity and collapses data from a variety of particle softnesses. We thus demonstrate how introducing heterogeneity to active matter leads to new emergent behaviors as well as a deeper theoretical understanding.

606111 High-Throughput Computational Screening of Metal-Organic Frameworks for Selective Adsorption of C_2H_6 over C_2H_4

Abstract:

C_2H_6/C_2H_4 separation is one of the most crucial processes in the petrochemical industry and currently practiced by energy-intensive distillation technology. As an alternative, adsorption-based separation is considered as technically feasible and economically viable, and there has been considerable interest to develop advanced adsorbents for C_2H_6/C_2H_4 separation. In this study, we computationally screen a large set (> 10000) of metal-organic frameworks (MOFs) to identify candidates for selective adsorption of C_2H_6 over C_2H_4. The quantitative relationships are established between adsorption performance metrics (C_2H_6 working capacity and C_2H_6/C_2H_4 selectivity) and structural descriptors (pore size and surface area), as well as isosteric heat. Top-performing 16 MOFs are identified with C_2H_6/C_2H_4 selectivity > 2.16 and C_2H_6 working capacity > 0.53 mol/kg, which are superior to the literature benchmarked MOF namely ZJU-120a. It is revealed that both pore size and pore shape are dominant factors to govern C_2H_6-selective adsorption. Moreover, the separation performance in MOF duplicates and geometrical analogues is elaborately analyzed and categorized. Finally, six strategies are proposed to boost C_2H_6/C_2H_4 separation by introducing catenation, aliphatic/aromatic linkers, pillar layers, framework flexibility, topological variants, and different metal nodes. These strategies provide bottom-up guidelines for the rational design of new MOFs for high-performance C_2H_6/C_2H_4 separation.

600270 Multi-Layer Connectivity-Based Atom Contribution (m-CBAC) Approach: Fast and Accurate Charge Assignments to MOFs for Adsorption Simulations

Abstract:

Metal-organic frameworks (MOFs) have shown great potential in a variety of energy-related applications such as gas separations and storage. With increasing interest in this emerging class of novel materials, molecular simulations can play a critical role in the efficient exploration of their adsorption properties and thus the MOF discovery. In these calculations, accurate charge assignments to the framework atoms are essential. To facilitate the large-scale computational simulation of MOFs, we develop a multi-layer connectivity-based atom contribution (m-CBAC) approach as an efficient, robust, and accurate approach for charge assignments. Different from the original connectivity-based atom contribution (CBAC) method, which uses 1st layer connectivity (i.e., directly bonded atoms) of a target atom, we expand on to incorporate multi-layer connectivity from the 0th layer up to the 2nd layer. This m-CBAC approach is developed with an automatic procedure of a supervised machine learning for classification, and it also allows further improvements along the time with more data to be included in the training set. In this study, an extensive set of ~2700 MOFs from the CoRE MOFs database with the density derived electrostatic and chemical (DDEC) charges are used to train the m-CBAC model. For the charge predictions, the m-CBAC approach assigns charges by successively searching from the highest-level layer (i.e., 2nd layer, presumably the most accurate) to the lowest-level connectivity (i.e., 0th layer, element type). With this approach, it becomes possible to assign charges for essentially all MOFs, with a high accuracy, and within seconds for a single structure. The m-CBAC approach is orders of magnitude computationally less expensive than quantum mechanical methods. The charges assigned using m-CBAC resemble the DDEC charges very well with a Pearson coefficient of as large as 0.988. The accurate charge assignments from m-CBAC are also reflected on their reliable predictions of CO_2 Henry coefficients in MOFs. Besides, a large-scale computation of the CO_2 Henry coefficients for ~12,000 MOFs in the recently released CoRE database has proved the capability of the m-CBAC approach. Overall, the m-CBAC approach can enable fast charge assignments for diverse MOFs with a good accuracy, and the software for m-CBAC charge assignments is made available along with this work.

602112 Incorporating Intrinsic Flexibility Effect into MOF Adsorption Properties Simulation at a Low Computational Cost

Abstract:

Computationally predicting the adsorption properties of Metal organic Frameworks (MOF) in a high throughput way has become a useful means of identifying promising materials for chemical separations. In almost all cases, high throughput simulations assume MOF structures as rigid for computational convenience. All MOFs, however, have intrinsic flexibility due to thermal vibration of atoms. Recent studies have suggested that this ubiquitous flexibility may have significant impacts on molecular adsorption in MOFs. Methods have been developed to predict the influence of this type of flexibility on MOF adsorption properties using molecular dynamics (MD) simulations, but these methods are computationally expensive and rely on the existence of reliable force fields.

We will describe a method of estimating the impact of intrinsic flexibility on MOF adsorption, which does not use any force fields for MOF framework. This approach uses experimentally available thermal vibration parameters to provide information about local flexibility. We show how this information can be used to reliably assess whether atomic vibrations make a significant impact on molecular adsorption in MOFs. Data from a variety of single component and binary mixtures adsorbed at dilute and non-dilute conditions forms our test sets. This method creates a useful way to establish the level of detail that is required for molecular modeling of adsorption in experimentally accessible MOFs to give quantitatively reliable results.

601234 Combining High-Throughput Molecular Simulations and Machine Learning to Optimize Adsorption Processes

Abstract:

High-throughput molecular simulations utilizing massively parallel high-performance computers allow now for the generation of an unprecedented amount of adsorption data. However, given the high dimensionality of state space (temperature, pressure, and composition) and/or the large number of porous materials governing the adsorption behavior, molecular simulation alone is not effective in finding optimal materials and conditions for adsorption processes. This talk will highlight the development of surrogate machine learning models that are trained on molecular simulation data and enable finding optimal material/condition combinations for gas storage or chemical separation applications: (a) desorptive drying for hydrogen-bonding solute/solvent mixtures in all-silica zeolites, (b) adsorption of BTEX mixtures onto zeolite nanosheets at membrane reactor conditions, and (c) gas storage in diverse porous materials.

596883 Computational Screening of Metal-Catecholate Functionalized Metal-Organic Frameworks for Room Temperature Hydrogen Storage

Abstract:

Hydrogen is a promising alternative to fossil fuels as it is non-toxic and carbon-free. However, due to its extremely low volumetric density, the storage and transport of hydrogen for practical applications remain a significant challenge, and high pressure and/or cryogenic temperature is required. Adsorption-based storage utilizing nanoporous adsorbents such as metal-organic frameworks (MOFs) can greatly reduce the storage pressure, but cryogenic or sub-ambient temperatures are required with current adsorbents, which limits the scope of applications. In this work, we aim to search for hydrogen-storage adsorbents that allow room temperature operation by looking at MOFs functionalized with metal catecholate groups, which have highly unsaturated open metal sites and thus greatly enhanced binding strength for hydrogen. We screened a dataset of 2736 zirconium-based MOFs that was constructed in a combinatorial fashion with wide varieties of topologies and linkers. By counting the possible sites that can be functionalized with metal catecholate groups, we were able to obtain the theoretical maximum hydrogen uptake in gravimetric and volumetric units for all of the MOFs. For the 100 MOFs with the highest theoretical uptakes, we built the functionalized structures on the computer and conducted grand canonical Monte Carlo (GCMC) simulations to predict the hydrogen uptake at adsorption (296 K, 100 bar) and desorption (296 K, 5 bar) conditions. The deliverable capacities were calculated by taking the difference of uptakes at those conditions, and we observed >5 wt% and >20 g/L deliverable capacities for some MOFs, which are very high for room temperature pressure swing adsorption cycles. Structure-activity relationships between the structural properties of MOFs and their hydrogen uptakes were also established.

611189 Pore Level Decomposition of Adsorption Isotherms in Metal Organic Frameworks

Abstract:

Metal Organic Frameworks (MOFs) demonstrate a variety of unique features due to their ability to selectively adsorb and retain guest molecules from gas and liquid phases. Adsorption properties are directly related to the specifics of the pore structure. A better understanding between the MOF structural parameters and adsorption properties is required for the guided design of novel MOFs and their modifications for advanced applications. MOFs are considered as nanoporous crystals processing regular 3-dimensional pore networks composed of pore compartments than differ in size, shape and chemistry. Here, we present a Monte Carlo (MC) simulation methodology for decomposition of the adsorption isotherm into the fingerprint isotherms, which reflect adsorption in individual pore compartments [1]. We consider two characteristic MOF structures, PCN-224 and ZIF-412, for which the distribution of adsorbate between the individual pore compartments was studied experimentally using adsorption crystallography based on *in-situ* XRD [2]. The pore network in PCN-224 represents a cubic network of alternating micropore channels and junctions of size 1.5 and 2.5 nm. The pore network in ZIF-412 has a complex micro-mesopore structure with three type, with pores of ca. 2.1, 2.5, and 3.9 nm. Employing MC simulation, we constructed the adsorption isotherms of three most common adsorbate probes (argon, carbon dioxide and nitrogen) in the ideal crystallographic models of PCN-224 and ZIF-412 and decomposed them into the fingerprint isotherms for the individual pore compartments. The results of simulation are found in qualitative agreement with the experimental data [2]. This work is motivated by the need of novel pore structure characterization methods, which account for the multiplicity of morphologies of different families of MOFs. Comparison of the reference fingerprint isotherms with the experimental adsorption isotherms allows to assess to what extend the structure of the practical sample differs from that of the ideal crystal due various types of defects, residual solvents, degree of hydration, pore blockage, etc.

This work is supported by the National Science Foundation (CBET grant no. 1834339)

1. Dantas, S.; Neimark, A. V., Coupling Structural and Adsorption Properties of Metal-Organic Frameworks: From Pore Size Distribution to Pore Type Distribution. *ACS Appl Mater Interfaces* **2020,** *12* (13), 15595-15605.
2. Cho, H. S.; Yang, J.; Gong, X.; Zhang, Y. B.; Momma, K.; Weckhuysen, B. M.; Deng, H.; Kang, J. K.; Yaghi, O. M.; Terasaki, O., Isotherms of individual pores by gas adsorption crystallography. *Nature Chemistry* **2019**.

602529 Modeling of Adsorption in Pristine and Defective Metal Organic Framework Materials

Abstract:

Metal-organic frameworks (MOFs) have attracted significant interest for adsorption due to their high degree of tailorability and large specific surface areas. The combination of tailorability and well-defined crystalline pores makes MOFs very promising candidates for highly selective adsorption.

Molecular-level modeling of adsorption and diffusion in MOFs almost always relies on models of MOFs that are defect-free (pristine). However, all real MOFs have defects, which affect adsorption by changing the types of pores within the MOFs. A fundamental understanding of how defects impact adsorption is important for identifying the limits of the performance of real material, developing improved design rules for new improved materials, and predicting and maximizing utilization of the material.

In this study, we focus on the study of how adsorption correlates with the number and types of defects. We initially consider UiO-66 with different levels of missing linker defects. The structures of the generated defective MOFs were optimized using periodic density functional theory with the CP2K simulation package. Adsorption isotherms were generated by carrying out grand canonical Monte Carlo (GCMC) simulations in Raspa. We investigated the effect of different adsorbate-adsorbent charge schemes by comparing isotherms with no framework charges and atomic charges calculated using DDEC6 and EQEQ methods. Different potentials for adsorbate were also tested for more accurate modelings.

610534 (Invited Talk) Mechanistic Insights into Deoxygenation Chemistry on Bimetallic Phosphides

Abstract:

Bimetallic catalysts are crystalline solids that consist of two different metals that can facilitate specific chemical reaction. The cooperative interactions between the two metals often result in higher stability, and functionality when compared to their monometallic counterparts. Adding non-metallic elements, such as phosphorous, allows for charge distribution between the metal and nonmetal atoms, which improves on these properties. Although bimetallic catalysts have been used in traditional chemical reactions, their application to biomass conversion is largely unexplored. In this presentation, I will share our work on deoxygenation of model aromatic compounds over different composition contained bimetallic catalysts ($M_XMo_{(2-X)}P$; M = Fe, Ru, Ni; $X = 0.8, 1.0, 1.2$) [3-5]. We explain our results in terms of binding energies, energetics along chemical transformation, and Bader charges.

608166 Ultrafast Hot Carrier Injection in Au/GaN: The Role of Band Bending and the Interface Band Structure

Abstract:

Plasmon photochemistry can potentially play a significant role in photocatalysis. To realize this potential, it is critical to enhance the plasmon excited hot carrier transfer and collection. However, the lack of atomistic understanding of the carrier transfer across the interface, makes it challenging to design more efficient system. In this work, we develop and apply the non-adiabatic molecular dynamics simulation to study hot carrier dynamics in the system of Au nanocluster on top of GaN surface. This method by including the electron-phonon interaction allows for a direct *ab initio* simulation of excited carrier dynamics in a large-scale system and for a long period of time. By setting up the initial excited hole in Au, the carrier transfer from Au to GaN is found to be on a sub-pico second time scale. After the hole has cooled down to the band edge of GaN, we find some of the charges can return back to Au. By applying different external potentials to mimic the Schottky-barrier band bending, the returning charge can be reduced effectively. Finally, with the understanding of the carrier transfer's pathway, we suggest that a ZnO layer between GaN and Au can effectively block the "cold" carrier from returning back to Au but still allow the hot carrier to transfer.

*This work is funded by the Joint Center for Artificial Photosynthesis, supported through DOE under Award number DE-SC0004993. NERSC and Oak Ridge Leadership Computing Facility provide computational support.

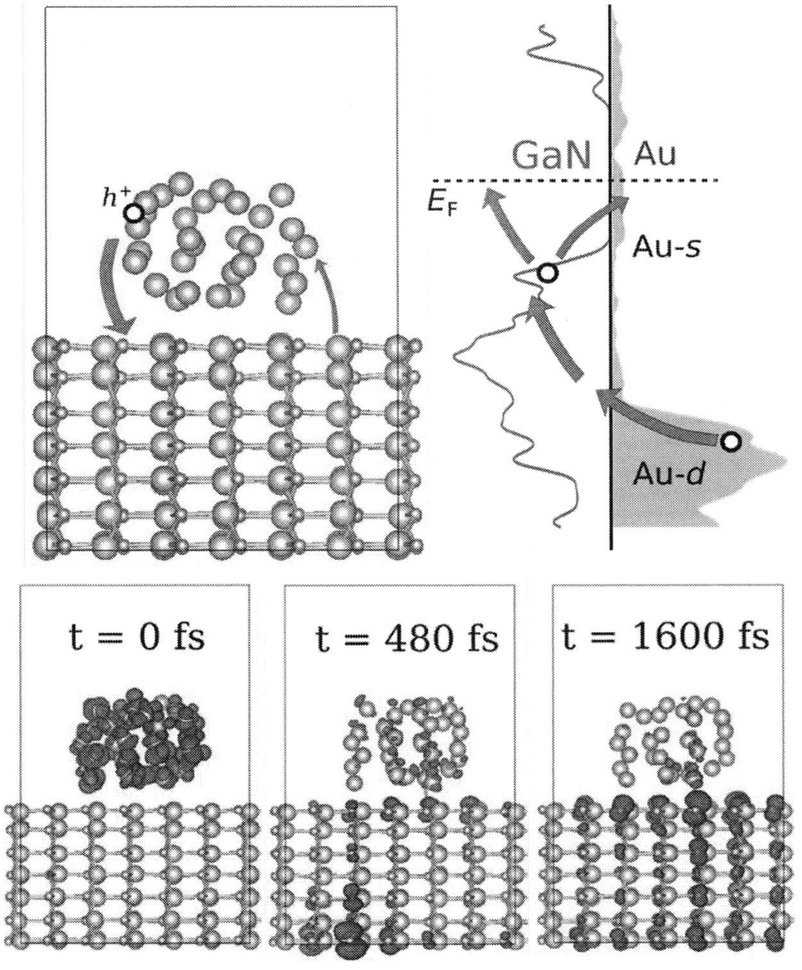

598436 Halides Induce Anisotropic Growth of Cu Nanocrystals

Abstract:

Properties of metal nanocrystals are generally controlled by their morphologies. Though Ag nanocrystals are used in commercial applications, the properties of Cu are comparable so that this earth-abundant material has drawn increasing attention. There have been many accounts of experimental, solution-phase syntheses of various Cu nanocrystals, including nanowires and nanoplates – however, the mechanisms of such anisotropic growth are still not fully understood.

It is commonly believed that capping agents, such as hexadecylamine (HDA) can induce anisotropic growth by selective binding to certain crystal facets. However, the emergence of round Cu nanoparticles synthesized in the presence of HDA does not seem to support this. Only when halide is present as an additive during nanocrystal synthesis are distinctive shapes observed. When Cl^- is the additive, nanowires form, but I^- leads to the formation of nanoplates. To elucidate the roles of halides in anisotropic Cu nanocrystal growth, we use density functional theory (DFT) to study the co-adsorption of halides and HDA on Cu surfaces with an analysis based on *ad initio* thermodynamics.

Both Cl^- and I^- are found to displace a chemisorbed HDA self-assembled monolayer (SAM) on Cu(100) and Cu(111). In the case of Cl^-, HDA SAMs are disrupted at lower Cl coverages on Cu(111) than on Cu(100). This can result in full protection for Cu(100) from Cu ion addition and a lack of protection for Cu(111) and is consistent with the experimentally observed formation of nanowires with long {100} side facets. *Ab initio* thermodynamics calculations show that, under the conditions for which nanowires are predicted and observed, introducing a trace amount of I^- can displace both Cl^- and HDA on both Cu surfaces, which creates a strong thermodynamic driving force for nanoplates with large {111} facets. The predicted evolution of Cu nanocrystal shape under different synthetic conditions agrees well with experimental results.

606718 First-Principles Study of Lithium-Air Batteries Based on Tri-Molybdenum Phosphide (Mo_3P) Nanoparticles

Abstract:

Lithium-air batteries are promising competitors to replace traditional lithium ion batteries due to the extremely high theoretical energy density. Despite extensive experimental research for decades, it is still a challenge to find a stable cathode material for both discharge (ORR) and charge (OER) reactions with a low overpotential. Here, we employ density-functional-theory (DFT) calculations to examine the mechanism, thermodynamic overpotential (η) and cyclability of a novel Mo_3P-based Lithium-air battery. We also compare and rationalize our findings with direct comparison with the recent experimental results of our collaborators. In this type of battery, Li_2O_2 is found as the only discharge product up to 1000 cycles due to the two-electron transfer reaction between Li^+ and O_2. Our DFT results show that the Mo-terminated Mo_3P (110) surface will be spontaneously oxidized in the oxygen-rich environment, thus forming a kinetically stable MoO layer to bring about the initial Li^+/e^- transfer from aprotic electrolyte to cathode. Furthermore, we find that the initial Li^+/e^- transfer reaction is the potential determining step (PDS) during ORR with a quite low DFT-based overpotential ($\eta = 0.06$ V), in agreement with our experimental detection ($\eta = 0.08$ V). In OER, the DFT-based overpotential was determined as 0.33V coupled with the desorption reaction of LiO_2 on the MoO layer covered catalyst, which is in agreement with experimental value ($\eta = 0.27$ V) but larger than that ($\eta = 0.12$ V)[1] on the kink sites of the crystal majority Li_2O_2 (0001) facet.[2,3] Therefore, we reveal that the MoO layer terminated Mo_3P cathode gives rise to a low-overpotential discharge reaction. However, the starting of charge process happens on the surface of agglomerated Li_2O_2 discharge product, while Mo_3P cathode becomes more and more important after most of the Li_2O_2 dissolved. Our results help shed fundamental insight on the role of ultrathin oxide monolayers on phosphide nanoparticles, and how this can potentially lead to a new nanomaterials design dimension.

1. Hummelshøj, J. S., Luntz, A. C. & Nørskov, J. K. Theoretical evidence for low kinetic overpotentials in Li-O_2 *Journal of Chemical Physics* **138**, 34703 (2013).

2. Radin, M. D., Tian, F. & Siegel, D. J. Electronic structure of Li_2O_2 (0001) surfaces. *Journal of Materials Science* **47**, 7564–7570 (2012).

3. Radin, M. D., Rodriguez, J. F., Tian, F. & Siegel, D. J. Lithium peroxide surfaces are metallic, while lithium oxide surfaces are not. *Journal of the American Chemical Society* **134**, 1093–1103 (2012).

610798 First-Principles Modeling of Thermal Degradation of CO_2-Loaded Aqueous Amine Solvents

Abstract:

Chemical absorption using an aqueous amine-based solvent appears to be the most promising near-term solution for post-combustion capture of CO_2 from flue gas.[1,2] Monoethanolamine (MEA) has been widely studied and is considered the benchmark solvent. Although it exhibits several advantages including a relatively high absorption rate, high solubility, low volatility, and low viscosity, widespread commercial implementation of MEA-based CO_2 capture systems remains hindered by costs associated with problems arising from solvent degradation and corrosion, in addition to the high parasitic energy consumption during regeneration.[1–8] Makeup of MEA loss from the solvent contributes about (~10 %) to the total cost of CO_2 capture, including the energy requirements for the reclamation process.[9–11] Moreover, some products of degradation may be corrosive and hazardous if released to the environment.[12,13]

Thermal degradation has typically been defined as degradation occurring due to elevated temperatures in the stripper, and tends to increase with CO_2 loading.[14,15] Thermal degradation of primary amines including MEA is thought to be initiated by cyclization of carbamates, which may result in the formation of 2-oxazolidinone (OZD) that is a minor product and proposed intermediate.[16] The OZD may undergo further reaction with amines to form diamines, imidazolidinones and urea, which are typically reported to be major degradation products.[14–16] Thermal degradation of CO_2-loaded aqueous amines could be mitigated by reducing the stripper operating temperature. However, lowering the temperature also typically increases the overall energy cost required for solvent regeneration.[17,18] The molecular mechanisms underlying thermal degradation of aqueous amines remain unclear, while an improved mechanistic understanding may provide valuable hints on how to improve the performance of existing solvents and in the rational design and synthesis of more degradation-resistant solvent materials.

In this talk, we will present our recent computational studies on the molecular mechanisms of thermal degradation of aqueous MEA in the CO_2 capture process. We have evaluated key elementary reactions using various computational methods. Static quantum mechanical calculations with an implicit solvent model were used to assess the relative stabilities of possible intermediates and products, and *ab initio* molecular dynamics (AIMD) simulations combined with metadynamics sampling were employed to predict free-energy barriers for key reaction steps, with particular focus on the effect of direct interaction between reaction intermediates and surrounding water molecules. Our work highlights that the thermal degradation mechanisms are largely governed by the competition of intramolecular and intermolecular interactions, besides steric constraints, which is a strong function of the arrangement and dynamics of water surrounding the intermediates.

References

[1] G. T. Rochelle, *Science*, 2009, **325**, 1652–4.

[2] J. D. Figueroa, T. Fout, S. Plasynski, H. McIlvried and R. D. Srivastava, *Int. J. Greenh. Gas*

Control, 2008, **2**, 9–20.

[3] U.S. Department of Energy, *Carbon Capture Technology Program Plan*, 2013.

[4] D. M. D'Alessandro, B. Smit and J. R. Long, *Angew. Chem. Int. Ed. Engl.*, 2010, **49**, 6058–82.

[5] B. R. Strazisar, R. R. Anderson and C. M. White, *Energy and Fuels*, 2003, **17**, 1034–1039.

[6] M. R. M. Abu-Zahra, J. P. M. Niederer, P. H. M. Feron and G. F. Versteeg, *Int. J. Greenh. Gas Control*, 2007, **1**, 135–142.

[7] M. R. M. Abu-Zahra, L. H. J. Schneiders, J. P. M. Niederer, P. H. M. Feron and G. F. Versteeg, *Int. J. Greenh. Gas Control*, 2007, **1**, 37–46.

[8] H. M. Stowe and G. S. Hwang, *Ind. Eng. Chem. Res.*, 2017, **56**, 6887–6899.

[9] A. B. Rao and E. S. Rubin, *Environ. Sci. Technol.*, 2002, **36**, 4467–4475.

[10] A. J. Reynolds, T. V. Verheyen, S. B. Adeloju, E. Meuleman and P. Feron, *Environ. Sci. Technol.*, 2012, **46**, 3643–3654.

[11] G. S. Goff and G. T. Rochelle, *Ind. Eng. Chem. Res.*, 2006, **45**, 2513–2521.

[12] G. S. Goff, 2005, 283.

[13] S. Chi and G. Rochelle, *Ind. Eng. Chem. Res.*, 2002, **41**, 4178–4186.

[14] J. Davis and G. Rochelle, *Energy Procedia*, 2009, **1**, 327–333.

[15] C. Gouedard, D. Picq, F. Launay and P. L. Carrette, *Int. J. Greenh. Gas Control*, 2012, **10**, 244–270.

[16] S. J. Vevelstad, I. Eide-Haugmo, E. F. Da Silva and H. F. Svendsen, *Energy Procedia*, 2011, **4**, 1608–1615.

[17] B. A. Oyenekan and G. T. Rochelle, *AIChE J.*, 2007, **53**, 3144–3154.

[18] G. T. Rochelle, *Curr. Opin. Chem. Eng.*, 2012, **1**, 183–190.

610484 Assessment of the Two-Phase Thermodynamic Method for Calculating Adsorption Isotherms

Abstract:

Adsorption isotherms are fundamental quantities of interest for understanding separation and catalysis phenomena. They are typically computed by grand-canonical or Gibbs ensemble Monte Carlo methods. These methods rely on potentially challenging particle insertion/deletion moves and may present a particular challenge for simulations using a first-principles potential energy surface. As an attractive alternative, the two-phase thermodynamic (2PT) model[1] allows one to calculate the free energy of a system from vanilla canonical molecular dynamics simulations. The 2PT method decomposes the total density of states into a gas-like component and a solid-like component. The former accounts for all fluidic effects and is treated as a hard-sphere gas. The solid-like component represents the non-zero, higher frequency modes and is handled as harmonic oscillators, which allows for straightforward quantum corrections by using the appropriate statistics. The 2PT method has been demonstrated to be capable of calculating the vapor-liquid coexisting properties and has recently been applied to study the free energy and stability of molecules adsorbed in carbon nanotubes[2] and zeolites[3]. In this talk, we will describe the development of the 2PT method for calculating adsorption isotherms and present the results for a united-atom methane model and the TIP4P water. We will discuss important caveats with the direct application of the original 2PT method and the modifications to the method that are necessary to obtain a correct description of adsorption equilibria.

(1) Lin, S.-T.; Blanco, M.; Goddard, W. A. The Two-Phase Model for Calculating Thermodynamic Properties of Liquids from Molecular Dynamics: Validation for the Phase Diagram of Lennard-Jones Fluids. *J. Chem. Phys.* **2003**, *119*, 11792–11805.

(2) Kumar, H.; Mukherjee, B.; Lin, S.-T.; Dasgupta, C.; Sood, A. K.; Maiti, P. K. Thermodynamics of Water Entry in Hydrophobic Channels of Carbon Nanotubes. *J. Chem. Phys.* **2011**, *134*, 124105.

(3) Bukowski, B. C.; Bates, J. S.; Gounder, R.; Greeley, J. Defect-Mediated Ordering of Condensed Water Structures in Microporous Zeolites. *Angew. Chem. Int. Ed.* **2019**, *58*, 16422–16426.

609349 Monte Carlo Simulations Exploring Adsorption at Liquid/Vapor, Liquid/Liquid, and Solid/Liquid Interfaces

Abstract:

This presentation will review applications of Monte Carlo simulations with enhanced sampling algorithms to investigate preferential adsorption and interfacial structure for liquid nanodroplets, coexisting fluid phases, and chromatographic systems.

597641 Minkowski Functionals for Phase Behavior Under Confinement

Abstract:

In this work, the Minkowski functionals are used as a framework to study how morphology (i.e. the shape of a structure) and topology (i.e. how different structures are connected) influence wall adsorption and capillary condensation under tight confinement. Numerical simulations based on classical density functional theory (DFT) are run for a wide variety of geometries using both hard-sphere and Lennard-Jones fluids. These DFT computations are compared to results obtained using the Minkowski functionals. It is found that the Minkowski functionals can provide a good description of the behavior of Lennard-Jones fluids down to small system sizes. In addition, through decomposition of the free energy, the Minkowski functionals provide a good framework to better understand what are the dominant contributions to the physics of a system. Lastly, while studying the phase envelope shift as a function of the Minkowski functionals it is found that topology has a different effect depending on whether the phase transition under consideration is a first- or a second-order transition.

610846 Understanding Amorphous Silica Coatings on the Platinum Surface with First Principles Calculations

Abstract:

Semi-permeable silica membranes are attractive as protective coatings for electrocatalysts, both for blocking contaminants from poisoning the catalyst and for preventing the aggregation of supported catalyst nanoparticles. Such coatings are not passive bystanders but can also affect the catalytic properties. However, to utilize the interplay of coating and catalyst, a better understanding of the coating/catalyst interface on the atomic scale is needed.

Here, we apply first principles calculations and ab initio molecular dynamics simulations to investigate how amorphous silica interacts with the surface of platinum metal electrocatalysts. Using a crystalline model interface structure, we compare different interface terminations and establish which bonds are formed between the SiO_2 membrane and the Pt(111) surface in aqueous electrolytes. We discuss how the dynamic nature of the interface may affect the catalytic properties of the SiO_2/Pt system.

The SiO_2/Pt system can be considered a simple model system for oxide-coated electrodes, and our approach can be readily extended to oxide coatings in other electrochemical devices such as batteries.

608248 Size-Selective Dynamics of Large Hetero-Epitaxial Islands on FCC(001) Surfaces

Abstract:

In this work, we use molecular dynamics simulations to investigate the diffusion of hexagonal silver monolayer islands on copper and nickel substrates. Our results indicate that certain sized islands diffuse orders of magnitude faster than other sized islands and even single atoms under similar conditions. An analysis of several low-energy pathways reveals two governing processes: a rapid, glide-centric process and a slower, vacancy-assisted one. We believe that the relative magnitude of the vacancy formation energies plays a role in determining the size-selective diffusion of these islands. These findings should provide insight to future experimental research on the size distribution and shapes observed during the growth of thin films in similar systems. Furthermore, our results should have implications for Monte Carlo models of film growth which generally allow only single atom diffusion, thus neglecting such complex, cooperative mechanisms.

608759 Investigating the Tribological Properties of Monolayer Films through High-Throughput Screening and Machine Learning

Abstract:

Monolayer films have shown promise as a potential solution for overcoming issues associated with friction and wear in nanoscale devices. Dense layers of surface-bound chains can prevent direct surface-surface contact, thus protecting the surface and reducing frictional forces. The tribological properties of these films have been shown to be closely related to their chemical composition [1, 2], presenting a near-infinite chemical parameter space that can be utilized to optimize tribological properties. In response, complex high-throughput computational screening studies have been enabled through the use of the Molecular Simulation and Design Framework (MoSDeF) [3] and the Signac framework [4]. MoSDeF is a Python-based software suite designed to set up chemical systems in a scriptable manner, while Signac provides a set of tools to create and manage a data workspace. Both software packages allow for a large parameter space to be efficiently explored through non-equilibrium molecular dynamics simulations. Here we examine the tribological effects of terminal group chemistry and film composition, encompassing approximately 36,000 distinct monolayer systems, by far the largest dataset of the tribological properties of monolayer films reported to date. To discover relationships hidden in this large volume of data, machine learning algorithms are employed. Specifically, building upon prior work [1], we use the RDKit library [5], a cheminformatics toolkit, along with the random forest algorithm implemented in scikit-learn [6] to develop a predictive model for the role of terminal group chemistry and film composition on tribological properties. Through this approach we are able to identify which aspects of terminal group chemistry most strongly impact the monolayer tribological properties. For example, it is observed that molecular shape most strongly influences the coefficient of friction, whereby linear and planar molecules are preferable, while charge distribution is most closely linked to adhesion. Compared to previous work, this project explores a much wider parameter space by extending the range of terminal groups examined, adding new backbone chemistries, and examining mixed monolayers. In addition to the random forest algorithm, we also consider other machine learning algorithms and compare their effectiveness in interpreting large volumes of data.

References

[1] Andrew Z. Summers, Justin B. Gilmer, Christopher R. Iacovella, Peter T. Cummings, and Clare McCabe. "MoSDeF, a Python Framework Enabling Large-Scale Computational Screening of Soft Matter: Application to Chemistry-Property Relationships in Lubricating Monolayer Films". Journal of Chemical Theory and Computation 2020 16 (3), 1779-1793. DOI: 10.1021/acs.jctc.9b01183

[2] Brewer, N. J.; Beake, B. D.; Leggett, G. J. Friction Force Microscopy of Self-Assembled Monolayers: Influence of Adsorbate Alkyl Chain Length, Terminal Group Chemistry, and Scan Velocity. Langmuir 2001, 17 (6), 1970– 1974, DOI: 10.1021/la001568o

[3] "MoSDeF" [Online]. Available: https://github.com/mosdef-hub.

[4] "signac" [Online]. Available: https://github.com/glotzerlab/signac.

[5] "RDKit Library" [Online]. Available: https://github.com/rdkit/rdkit.

[6] "Scikit-learn" [Online]. Available: https://github.com/scikit-learn/scikit-learn

606902 Computational Investigations of the Effect of Electrostatic Environment on Silica Self-Assembly

Abstract:

While looking to Nature for synthetic inspiration is a useful practice, the complexity of such synthetic routes can make laboratory-based mimics difficult to rationally design. Computational modelling can be used to understand on a fundamental level how aspects of reaction environment cause certain reactions to occur. In particular, this work focuses on the self-assembly of silicic acid into silica oligomers, a process that is conducted by the protein silaffin to generate complex exoskeletons in marine organisms with impressive morphological control. While work has been done to understand the behavior of silica in "bio-inspired" environments (e.g., using a surface of peptides on which to grow silica), relatively little research has been done to understand how simple charged interfaces affect the condensation of silica. What has been shown thus far is that minor changes in electrostatic environment can significantly affect the morphology of the resultant silica structures. To probe this effect on a fundamental level, static DFT and molecular dynamics simulations in the gas phase and solvent were conducted on an idealized model system to determine (1) the effect of a charged interface on the energetics of the reaction network and (2) the dynamic behavior of the solution that leads to particular silica morphology. Understanding the effects of electrostatic environment on silica oligomerization in isolation could lead to better synthetic control in laboratory settings.

607490 Ultrafast Melting at Metal-Semiconductor Interfaces: A Molecular Dynamics Study

Abstract:

Quantifying the thermodynamics of interfacial melting is important for several applications such as laser machining and designing radiation hardened electronic devices. Due to the short time and length scales that are involved in the process, molecular simulations provide an ideal path for understating the interfacial melting phenomena. In this work we investigate tungsten (W) – silicon carbide (SiC) interfaces which are relevant in modern high-power electronic devices. Classical molecular dynamics simulations along with the two-temperature model (TTM) were used to study the radiation induced melting of the interface while a neural network-based machine learning framework was used to track the transient re-crystallization process. The damage endured by the semiconductor at the interface following the radiation damage is discussed in detail while the re-crystallization of the metal is described in terms of the classical nucleation theory.

599736 Can We Define a Unique Microscopic Pressure in Inhomogeneous Fluids?

Abstract:

The estimation of the microscopic pressure in an adsorbed thin film on a planar surface remains a challenge in both experiment and theory. While the normal pressure is well-defined for a planar surface, the tangential pressure parallel to the surface at a point is not uniquely defined at the nanoscale. We show that by integrating the local tangential pressure over a small region of space, Δz, roughly the range of the intermolecular forces, it is possible to define an effective tangential pressure that is unique and free from ambiguities in the definition of the 'virial-route' local tangential pressure. Grand canonical Monte Carlo simulation results suggest that this characteristic length Δz for the adsorbed contact layer next to the wall has a weak correlation with the adsorbate-solid interaction strength, and the value of Δz converges for large slit pores. With that, we can also define an effective thickness for the adsorbed layer, the value of which agrees with the reported data. This unequivocal definition of the effective tangential pressure would help bridge the gap between experiment and theory in comparing and estimating the microscopic pressures in highly inhomogeneous systems.

611769 Computational Design of Stable Hybrid Perovskite Systems: From Mixed Cations to Surface Passivation and Elastic Properties

Abstract:

Here, we discuss how density functional theory (DFT) methods can help us understand the stability and efficiency challenges for hybrid organic-inorganic perovskite (HOIP) systems. Recently, we have been studying the fundamental chemistry behind the moisture degradation mechanism of 3D hybrid perovskites.[1] We have now expanded this work to study layered quasi-2D systems. A fundamental understanding of chemical interactions in these hybrid systems, garnered by a range of DFT-based computations, allows us to put forth strategies to help improve stability and efficiency of a range of HOIP-based devices. Few examples of these strategies discussed here include: mixed short organic cations, supramolecular interactions between long cations, surface passivation with acid and amine ligands and the use of defect-healing agents. We have applied these design strategies to e.g., create better efficiency LED devices. We also discuss the role of fundamental chemistry and interactions between different organic and inorganic constituents in dictating the mechanical and elastic properties of such HOIP systems, as we have discussed in a recent article.[2] It is elucidated how a multipronged computational approach which studies a range of physical and chemical phenomena from elastic properties to water intercalation, dimensional engineering and cations mixing, can create a comprehensive framework for optimization and design of this complex class of hybrid systems.

[1] Kakekhani, A., Katti, R. N., & Rappe, A. M. (2019). Water in hybrid perovskites: Bulk MAPbI3 degradation via super-hydrous state. *APL Materials*, *7*(4), 041112.

[2] Reyes-Martinez, M. A., P. Tan, A. Kakekhani, S. Banerjee, A. A. Zhumekenov, W. Peng, O. M. Bakr, A. M. Rappe, and Y. L. Loo. "Unraveling the Elastic Properties of (Quasi) Two-Dimensional Hybrid Perovskites: A Joint Experimental and Theoretical Study." *ACS Applied Materials & Interfaces* (2020).

610831 Prediction of Adsorption Thermodynamics in MOFs and Cofs Via Ensemble Learning

Abstract:

We discuss how ensemble learning (EL) allows for a rapid prediction of the partition function for fluids adsorbed in nanoporous materials. For this purpose, we carry out flat histogram simulations for the adsorption process and train EL models on the simulated datasets. Once partition functions are predicted by the EL models, all thermodynamic properties of adsorption can be rapidly determined for a wide range of conditions, without any additional simulations. We apply various EL approaches to carbon capture and storage, hydrogen storage and gas separation in MOFs and COFs. In particular, we find that the diversity ensemble learning approach, obtained from neural networks with different types of architectures, provides the most accurate predictions, when compared to the experiment and to Expanded Wang-Landau simulation results. The EL models so obtained give access to the materials selectivity in the case of the dsorption of mixtures and to a number of key thermodynamic properties, such as the desorption free energy which measures the energetic cost to regenerate the adsorbent.

606275 Message Passing Neural Networks for Partial Charge Assignment in Metal-Organic Frameworks

Abstract:

Virtual screenings of thousands of metal-organic frameworks (MOFs) can accelerate and reduce the cost of discovering MOFs for their applications in gas storage, separation/purification, and sensing. In molecular mechanics simulations of gas adsorption/diffusion in MOFs, the adsorbate-MOF electrostatic interaction is typically modeled by placing partial charges on the atoms of the MOF. It is critical to develop computationally inexpensive methods to assign atomic partial charges to a MOF that accurately reproduce the electrostatic potential in its pores. Herein, we design and train a message passing neural network to predict the atomic partial charges on MOFs under a charge neutral constraint. A set of ca. 2266 MOFs labeled with high-fidelity partial charges, derived from electronic structure calculations, serve as training examples. In an end-to-end manner, from the crystal graph representing the MOF, our message passing neural network machine-learns features of the local bonding environments of each atom and learns to predict partial charges from these features. Our trained message passing neural network can assign high-fidelity partial point charges to MOFs with orders of magnitude lower computational cost, thereby enhancing the accuracy of virtual screenings of MOFs for their adsorption-based applications.

602469 High-Throughput Discovery of Metal-Organic Frameworks for Cooperative CO_2 Adsorption

Abstract:

Recently, a new class of post-synthetically modified metal-organic frameworks (MOFs) with non-Langmiur stepped isotherms have been discovered and tuned to reversibly and selectively adsorb CO2 under common flue gas conditions. However, very few MOFs are known to exhibit step-like isotherms, a result of a cooperative adsorption phenomenon. Here, we present a computational screening procedure to discover new CO2 adsorbent MOFs with the potential for step-like isotherms and cooperative adsorption. Our workflow is based on the hypothesis that the distance between accessible, undercoordinated metal sites is a key indicator for whether a MOF modified with ethylenediamine will exhibit cooperative adsorption. We screen the Computational-Ready Experimental MOF (CoRE-MOF) database using the fast marching algorithm to assess metal site distances given arbitrary pore geometries, and discuss candidate materials for experimental validation.

600720 Gaussian Process-Based Model Order Reduction for the Prediction of Gaseous Storage in Metal-Organic Frameworks.

Abstract:

Metal-organic frameworks (MOFs) are a unique class of crystalline porous materials with potential applications in a multitude of industrial settings including gas storage. The nature of MOFs means that their structures are highly tailorable allowing a large number to be synthesized. This poses the inherent problem of identifying promising MOFs for a given application without the need for complex and time-consuming calculations.

This study aims to further the understanding of MOFs by deriving a novel method that uses synthesised MOFs to develop a prediction technique that results in faster and more efficient material design and discovery for gas storage applications. The ability to accurately predict gas adsorption properties and identifying the most dominant geometrical features of MOFs affecting adsorption phenomena, allows scientists to quickly and efficiently design and discover new MOFs. To do this, we adopt a Gaussian Process (GP) surrogate modelling approach, applied to a large database of MOFs that have been experimentally synthesized. This GP surrogate is then used to perform a Global Sensitivity Analysis of the data, to quantify both the most important geometric characteristics of MOFs and the interaction between them, which is also then used as part of a dimensionality reduction approach to identify the parameter subspace required to accurately predict gas adsorption. A comparison is then made between these subspaces for the different adsorbed gases to assess the ability to predict across different gas species, here oxygen and methane.

Results, shown in Figure 1, demonstrate that the surface area and void fraction dominate the deliverable capacity of both oxygen and methane in MOFs. The dimensionality reduction shows that the pore properties can be confidently reduced to just three primary dimensions while also improving the prediction accuracy when the storage gas is varied. Overall, the use of a GP surrogate model for a Global Sensitivity Analysis and a reduced order model has revealed the accurate determination of important pore properties and has helped pave the way for accurate prediction of MOF performance regardless of the gas. This will aid experimental researchers in designing MOFs more efficiently concentrating on optimising the important pore properties.

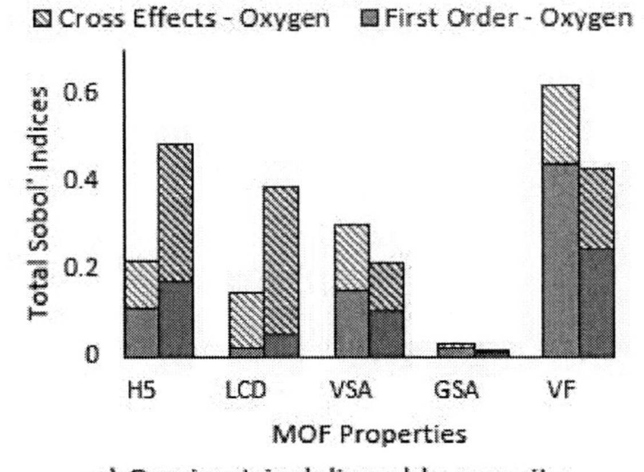

a) Gravimetric deliverable capacity

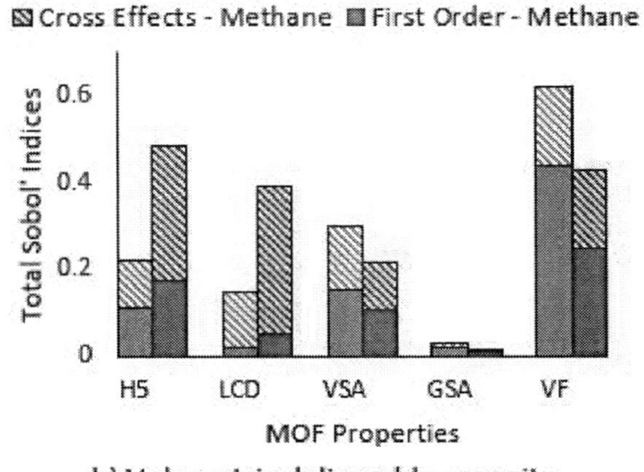

b) Volumetric deliverable capacity

Figure 1: The bar charts showing the calculated total Sobol' indices for each geometric feature with respect to the deliverable capacity of MOFs.

596692 A Database with Automated Quantum Chemistry Calculations and Machine Learning for Functional Transition Metal Complex Discovery

Abstract:

Large-scale databases of density functional theory (DFT) properties have proven invaluable in accelerating materials discovery, with notable example databases in both the solid state and small organic molecules. A similar database of transition metal (TM) complexes with DFT properties, however, is absent due to unique challenges in structure curation and domain of application of DFT for their properties. To address these challenges, we have developed open-source toolkits molSimplify and AutomaticDesign that incorporate both chemical rules and data-driven machine-learning (ML) models. Here, we highlight the first large-scale computational database of TM complexes, containing over 40,000 attempts of DFT geometry relaxation of octahedral complexes. From this database, we select a subset of over 24,000 complexes with higher-quality geometry and electronic structure. For these complexes, we calculated properties such as solubility, spin-splitting energy, redox potential, frontier orbital energies, and energetic functional dependence. To extend the number of complexes in the database continuously over time, we further built a workflow to automatically perform DFT geometry relaxations on diversity search-selected experimentally-determined complexes (i.e., from the Cambridge Structure Database) when supercomputing resources are idle. We also describe our automatic ML-model retraining module, which updates all our ML models when new results become available. With this automated workflow, we can readily monitor the evolution and quality of our ML models as the database grows. We anticipate our database, enhanced by automated quantum chemistry calculations and ML model retraining, to be a useful tool in accelerating discovery of functional transition metal complexes.

598384 Do Machine-Learned Formation Energies Enable Accurate Predictions of Compound Stabilities?

Abstract:

Machine learning (ML) models in chemistry and materials science are often tasked with predicting compound formation energies. These formation energies can then be used as inputs to construct phase diagrams and determine the stability of materials. Recent efforts have shown that ML models trained on density functional theory (DFT) calculated formation energies for solid-state materials achieve accuracies comparable to DFT itself (relative to experiment). In principle, these ML models should therefore be able to predict the stability of materials with comparable accuracy to DFT at a miniscule fraction of the computational expense.

In this work, we designed a set of tests to assess whether seven recently published machine learning formation energy models can accurately predict the stability of compounds. These tests include the reconstruction of ~20,000 phase diagrams available in the Materials Project database in addition to a simulated materials discovery problem, where the models are tasked with re-discovering stable compounds in sparse chemical spaces. Our findings suggest that ML models based only on chemical composition are generally not capable of reliably predicting compound stability and sheds insights into why that is. We do show that stability predictions can be improved by the incorporation of structure in the material representation.

In total, this work emphasizes the importance of testing machine learning models on real-world applications to assess their potential for incorporation in the materials discovery pipeline. We provide a set of publicly available tests to help facilitate this effort for future model development.

Preprint – C. Bartel, A. Trewartha, Q. Wang, A. Dunn, A. Jain, G. Ceder, 2020, *arXiv* 2001.10591

Code – https://github.com/CJBartel/TestStabilityML

605068 Physically Informed Deep Learning for Accelerated Photosensitizer Discovery

Abstract:

Photodynamic therapy is arising as a non-invasive treatment modality for cancer and other diseases. One of the key factors to determine the therapeutic function is the efficiency of photosensitizers (PSs). According to the latest studies, reducing the energy gap (ΔE_{ST}) between the lowest excited singlet state (S_1) and lowest excited triplet state (T_1) was an effective approach to yield highly efficient PSs and ΔE_{ST} could be fine-tuned through precise molecular donor-acceptor engineering. There is a growing literature where small families of small-ΔE_{ST} PS compounds are explored experimentally, or purely computationally at a slightly larger scale. Here, we report a large-scale data-driven search for novel PSs, with a special focus on red PS. This work employed a collaborative approach encompassing computational quantum chemistry, machine learning, organic synthesis, and nanoagent fabrication and testing.

By providing the list of commonly used donors, acceptors and bridges, there is a chemical design space of more than 7 million possible PSs. This is a vast search space that cannot be explored by chemical intuitions or Edison's approaches. Even performing quantum calculations of this vast space can take infeasibly long time. A data-driven approach by screening the ΔE_{ST} of 7 million PSs with a well-trained deep learning model provides a possible solution. Herein, we have created a high-throughput quantum calculation toolset for the construction of a novel dataset of 10k PSs with ΔE_{ST} information. This training dataset contains representative donors, acceptors and bridges with both Donor-Acceptor and Donor-Acceptor-Donor structures. A molecular graph convolutional neural network with molecular structure as features is trained on this dataset. In combination with a high-throughput PS generator, we could efficiently predict the ΔE_{ST} of possible PSs within the chemically intuited design framework. In addition to this physically informed design model, we have utilized a junction tree variational autoencoder as another deep learning method for the direct generation of possible PS candidates. These two approaches will be compared with results discussions. Furthermore, the experimental validation of chosen PSs with low ΔE_{ST} is conducted and presented. Finally the potentials and limitations of this work will be discussed.

610293 Quantum Chemistry-Informed Active Learning to Accelerate the Design and Discovery of Sustainable Energy Storage Materials

Abstract:

We employed Density Functional Theory (DFT) to compute oxidation potentials of 1,400 homobenzylic ether molecules to search for the ideal redoxmer design for non-aqueous flow batteries. The generated data were used to construct an active learning model based on Bayesian optimization that targets candidates with desired oxidation potentials utilizing only a minimal number of DFT calculations. The active learning model demonstrated not only significant efficiency improvement over the random selection approach but also robust capability in identifying desired candidates in an untested set of 112,000 homobenzylic ether molecules. Our findings highlight the efficacy of quantum chemistry-informed active learning to accelerate the discovery of materials with targeted properties from a vast chemical space.

598630 Color As a Source of Information in Liquid Crystal Sensors

Abstract:

Understanding the physical alignment of liquid crystals (LCs) near an interface is crucial for the design of electro-optical devices (e.g., liquid crystal displays [1]), nanoscale materials (e.g., semiconducting nanorods [2, 3]), and sensors [4, 5]. In this work, we are specifically interested in LC-solid interfaces, which provide a flexible platform for the development of sensors to detect air contaminants [6]. The working principle of such sensors relies on the alignment of LC molecules at interfaces, which can be controlled via tailored surface chemistries (e.g., metals and metal oxides) [6]. LC alignment and the evolution of such alignment can be visualized using orthoscopic polarized light microscopy [7]. This technique probes an LC sample with a beam of near-parallel rays of polarized light; LC alignment at the interface generates optical fields with different features (e.g., color and spatial textures) [4, 6, 8]. It has been recently shown that the features of these optical fields can be decoded by machine learning techniques to detect different types of air contaminants and to measure their concentration [13]. These studies also found that color plays a key role in distinguishing contaminant type and concentration.

In this work, we investigate color analysis techniques to transform macroscopic optical fields to physical parameters of interest such as the zenithal (tilt) and azimuthal (twist) angles of the LCs at the interface. This information is used to obtain insights into the physical origins of information in color fields. Alignment studies typically use a Berek compensator, which can quantify the optical retardance caused by birefringence (the difference between direction-dependent refractive indices) between the top and bottom interfaces of the sensor. Miller et. al. showed that LC alignment at the interface can be approximated using a Berek compensator [7]. However, a Berek compensator cannot measure optical retardance in a scalable manner. Specifically, it can not measure the evolution of the optical retardance for multiple samples simultaneously. A Michel-Levy chart is typically used to correlate color with optical retardance [10]. Sørensen et. al. formulated an equation to construct Michel-Levy charts quantitively [11] but they did not provide a direct equation to map color to LC alignment. Moreover, standard Michel-Levy charts do not consider the effect of different light sources. For instance, an approximation of optical retardance from samples under halogen light bulbs will not be accurate based on the standard Michel-Levy Chart.

In this work, we present a mathematical framework to map color in optical fields to LC alignments at the surface (and viceversa). The framework takes into consideration various aspects of relevance in LC-based sensor design (e.g., film thickness and light sources). We use our framework to construct a generalized Michel-Levy chart that captures optical retardance under different light sources. These capabilities allow us to analyze the effect of different light sources (e.g., halogen light vs. LED) and color space representations (e.g., RGB and LAB) on machine learning tasks. Specifically, we construct a 3D convolutional neural network to predict chemical species and concentrations from a temporal evolution of LC optical fields with varied light sources and color representations. This framework provides insights on how to co-design instrumentation (light sources) and the LC interface (film thickness) to extract maximum information and provides insights on the physical origins of information in LC optical responses. We use attribution methods (e.g., saliency maps [12]) to get insights of the governing dynamics of LC alignments.

[1] Drzaic, Paul S. *Liquid crystal dispersions*. Vol. 1. World Scientific, 1995.

[2] Karanikolos, Georgios N., et al. "Templated synthesis of ZnSe nanostructures using lyotropic liquid crystals." *Nanotechnology* 16.10 (2005): 2372.

[3] Yang, Chung-Sung, David D. Awschalom, and Galen D. Stucky. "Growth of CdS nanorods in nonionic amphiphilic triblock copolymer systems." *Chemistry of Materials* 14.3 (2002): 1277-1284.

[4] Bai, Yiqun, and Nicholas L. Abbott. "Recent advances in colloidal and interfacial phenomena involving liquid crystals." *Langmuir* 27.10 (2011): 5719-5738.

[5] Lowe, Aaron M., and Nicholas L. Abbott. "Liquid crystalline materials for biological applications." *Chemistry of Materials* 24.5 (2012): 746-758.

[6] Yu, Huaizhe, et al. "Amplification of Elementary Surface Reaction Steps on Transition Metal Surfaces using Liquid Crystals: Dissociative Adsorption and Dehydrogenation." *Journal of the American Chemical Society* 141.40 (2019): 16003-16013.

[7] Miller, Daniel S., et al. "Introduction to optical methods for characterizing liquid crystals at interfaces." *Langmuir* 29.10 (2013): 3154-3169.

[8] Lockwood, Nathan A., and Nicholas L. Abbott. "Self-assembly of surfactants and phospholipids at interfaces between aqueous phases and thermotropic liquid crystals." *Current Opinion in Colloid & Interface Science* 10.3-4 (2005): 111-120.

[9] Collings, Peter J. *Liquid crystals: nature's delicate phase of matter*. Princeton University Press, 2002.

[10] Spring, K. R., M. J. Parry-Hill, and M. W. Davidson. "Michel-Levy Birefringence Chart." *Olympus Microscopy Resource Center* (2010).

[11] Sørensen, Bjørn Eske. "A revised Michel-Lévy interference colour chart based on first-principles calculations." *European Journal of Mineralogy* 25.1 (2013): 5-10.

[12] Simonyan, Karen, Andrea Vedaldi, and Andrew Zisserman. "Deep inside convolutional networks: Visualising image classification models and saliency maps." *arXiv preprint arXiv:1312.6034* (2013).

[13] Cao, Y., Yu, H., Abbott, N. L., & Zavala, V. M. (2018). Machine Learning Algorithms for Liquid Crystal-Based Sensors. *ACS sensors*, *3*(11), 2237-2245.

600728 A Deep-Learning Potential for Crystalline and Amorphous Li-Si Alloys

Abstract:

This work investigates the ability of the deep-learning potential (DP) to describe structural, dynamic and energetic properties of crystalline and amorphous Li-Si alloys. Li-Si systems play an important role in the development of high energy lithium ion batteries. One challenge in simulating Li-Si systems is to balance the proper description of complex Li-Si interactions and the system size. Machine-learning potential paves an avenue to achieve this balance by describing complex interactions using machine-learning models. The machine-learning models enable us to investigate complex systems beyond the capability of the classical force fields using molecular dynamics simulations. We develop a DP for Li-Si systems with Li/Si ranging from 0 to 4.2 based on a vast dataset generated using the quantum mechanical calculations in an active learning procedure. Then we investigate the structural and dynamic properties of several crystalline and amorphous Li-Si systems using this developed DP. The DP can predict bulk densities, the radial distribution functions and diffusivity of Li in amorphous Li-Si systems at a level similar to the quantum mechanical calculations, while the molecular dynamics simulations are 20 times faster than the quantum mechanical calculations. We also discuss several issues when developing DP.

607895 Accelerating *Ab Intio*-Based Free Energy Sampling with Machine Learning

Abstract:

Ab initio Molecular Dynamics (AIMD) simulations are powerful for studying quantum mechanical phenomena, including chemical reactions, diffusion, and defect generations in materials, which require breaking and formation of bonds. Understanding not only the kinetics but also the thermodynamic properties of materials often require combining molecular dynamics with enhanced sampling techniques to capture the free energy landscape of the underlying phenomena. Unfortunately, free energy sampling with AIMD has been infeasible for large complex systems due to the computational cost of electronic structure calculations that are needed for each point in the configurational space. In this talk, we propose two approaches to overcome such bottleneck by using machine learning, namely the artificial neural network: (1) active learning of free energy surface during the adaptive biasing simulations and (2) on-the-fly learning of both *ab initio* force field and free energy surfaces during biased AIMD simulations. We demonstrate these two methods by applying them to catalytic reactions and defect migrations in covalently bonded solids.

597631 Charting the Design Space of Chemically-Heterogeneous Surfaces That Manipulate Water-Mediated Interactions

Abstract:

Water-mediated interactions constitute fundamental driving forces in a profound range of synthetic and natural materials. While such interactions have now been studied extensively at a theoretical level for the simplest of solutes with well-defined chemistries (e.g., ideal hydrophobes, simple ions, model colloids, etc.), the picture for highly heterogeneous solutes and surfaces that present multiple chemistries and interaction types remains far from well-understood. In this talk, I discuss the application of molecular simulations that are coupled to optimization algorithms to systematically explore and characterize the enormous space of chemically heterogeneous surfaces, particularly ones that balance hydrophobic and hydrophilic content. Our recent work addresses surfaces that systematically manipulate both solute affinity and transport properties, the latter including equilibrium (diffusive) as well as non-equilibrium (convective) behavior.

607666 Predicting Perturbations to Water Structure By Chemically Heterogeneous Self-Assembled Monolayers to Understand Context-Dependent Hydrophobicity

Abstract:

Predicting hydrophobic interactions between chemically heterogeneous surfaces (*i.e.*, surfaces with polar and nonpolar functional groups in close proximity) is a significant challenge due to cooperative and competing effects by the specific chemical properties of surface functional groups. These effects have led to hydrophobic interactions being labeled as context-dependent interactions, meaning that both the specific chemical groups and their spatial relation to one another dictate the strength of hydrophobic interactions. Extensive experimental and theoretical efforts have focused on understanding the underlying relationship between the chemical properties, interfacial water structure, and hydrophobic interactions. Recently, experiments using an atomic force microscope (AFM) measured the hydrophobic force between self-assembled monolayers (SAMs) with mixed nonpolar and polar functional groups, creating chemically heterogeneous surfaces. This study found that amide functional groups reduced the magnitude of hydrophobic interactions to a greater extent than amine functional groups. Understanding how these specific chemical groups uniquely perturb the surrounding interfacial water molecules would inform how these groups change the hydrophobicity of the material.

In this study, we utilized classical molecular dynamics (MD) simulations to investigate how amine, amide and hydroxyl functional groups influence the hydrophobicity of mixed SAMs. We used umbrella sampling to directly measure hydrophobic forces between mixed SAMs with compositions that mimic the experiments. Using this technique, we reproduced the qualitative trends of the AFM experiments. We further applied indirect umbrella sampling to measure the hydrophobicity of mixed SAMs with the same functional groups but different surface patterns. We then leveraged the atomistic resolution of MD to relate the changes in hydrophobicity to perturbations in interfacial water structure caused by the specific chemical groups decorating the SAMs. We were able to relate changes in the tetrahedral behavior and hydrogen bond network of the water molecules at the SAM interface to changes of hydrophobicity. This new understanding describing how specific chemical groups perturb interfacial water structure provides important information about how materials tune hydrophobic interactions and could be used to inform the design of new materials.

611198 Effect of Multivalent Surface Cations on Heterogeneous Freezing of Water

Abstract:

Heterogeneous ice nucleation is a crucial phenomenon in various fields of fundamental and applied science. In this study we investigate the effect of surface cations on freezing of water on muscovite mica. The exposed ion on mica's surface can be exchanged without affecting other properties such as surface roughness. We investigate freezing of water on natural (K+) mica and mica surfaces where K^+ is for Al^{3+}, Mg^{2+}, Ca^{2+}, and Sr^{2+}. We mica surfaces with ions of higher valency exposed are better ice nucleators; i.e. nucleation on Al^{3+}-mica surface occurs at higher temperatures than the mica surfaces with divalent ions. Molecular dynamics simulations indicate that multivalent cations enhance the probability of forming larger clusters of hydrogen bonded water molecules. The cations provide anchors for these clusters. These clusters could provide seedbed for ice nuclei. This represents a novel mechanism through which ions can enhance ice nucleation. The results of the combined experimental and simulation studies shed new light on the influence of surface ions on heterogeneous ice nucleation.

606145 Understanding the Aqueous Solubility and Adsorption Behavior of Surfactant Molecules on Metal-Water Interfaces Using Atomistic Simulations

Abstract:

Adsorption of surfactant molecules on metal surfaces is useful in many applications, such as corrosion inhibition, and synthesis of anisotropic metal nanoparticles, etc. We have studied the adsorption free energy of cationic and uncharged surfactants in different aggregation states via all-atom molecular dynamics simulations. In the un-aggregated state, both types of surfactants adsorb strongly without experiencing any free energy barrier. Adsorption of micelles is, however, different for cationic and uncharged surfactants. Cationic micelles experience a long-range repulsion from the metal surface because of the presence of a corona of counterions and hydration layer around them, which gets disturbed as the micelles approach the surface. Uncharged micelles, on the other hand, undergo a barrier-less adsorption because they lack a corona of counter-ions. Upon adsorption, both types of micelles slowly disintegrate in the timescale of 200 nanoseconds to strongly adsorb on the metal surface.

Along with the direct adsorption onto the metal surface, high aqueous solubility is another desirable characteristic of the surfactants for different applications. We have performed free energy calculations at the air-water interfaces to study the hydration free energy of surfactant molecules. We propose that an addition of a polar moiety, such as a hydroxyl group, to the terminal end of the alkyl tail improves the solubility of the surfactants by three orders of magnitude.

599274 Molecular Dynamics Simulations of Liquid-Liquid Interfacial Properties between Water and Hydrophobic Deep Eutectic Solvents

Abstract:

Deep eutectic solvents (DES) emerged recently as candidates for extracting organic molecules from aqueous solutions. The liquid-liquid interfaces between aqueous solutions and DES play a determining role in the ability of DES to extract organic molecules. A better understanding of the DES/aqueous liquid-liquid interfaces can help develop an efficient and cost-effective extraction process. We investigate the structural and dynamic properties of molecules at the liquid-liquid interfaces in several hydrophobic DES/water systems using molecular dynamics simulations. The analysed properties include density distribution and diffusion of DES composition and water molecules, and hydrogen bonding between DES compositions and water molecules at the interface. The simulation results enable us to develop a fundamental relationship between DES compositions and the properties of the DES-aqueous interface. This relationship can help us to develop DES that can form a desired interface with the aqueous solution.

604924 Star Polymers and Their Interaction with Lipid Membranes Via Coarse-Grained Molecular Simulations

Abstract:

Star polymers (SPs) of fifteen different architectures/ compositions were simulated both in solution and interacting with a model lipid membrane using a MARTINI coarse-grained model. These simulations yielded information on the equilibrium structure-property relationships of such macromolecules and the energetics of active membrane transport. Star polymers exist in solution in a nearly spherical shape and their size depends on both the total length of their polymer arms and the degree of hydrophobicity, with variation of ~30% observed in the radius of gyration between a fully hydrophilic SP and a fully hydrophobic SP. These properties were also measured during membrane translocation and SPs presented deformations of up to 40% or 100% depending on the length of the polymers arms. Such large deviations in size and shape could ultimately lead to changes in the drug loading/ release behavior. The deformations of the membrane were also recorded as was the energy barrier to transport which varied up to 100% depending on the hydrophobicity of the SP. The extraction of lipid molecules into the SP post-translocation was also observed, serving as a further proof-of-concept for the potential of these amphiphilic macromolecules to act as nanocarriers.

598894 The Influence of Sequenced Peptoids in Controlling Five-Fold Twinned Au Nanostar Formation

Abstract:

Five-fold twinned metal nanocrystals have attracted much attention due to their unique plasmonic properties, and a collection of five-fold twinned nanostructures have been synthesized in solution for Au, Ag and Cu, with the aid of chemical additives. It has been widely demonstrated that these chemical additives are critical in achieving such shape-controlled syntheses, and their roles may vary significantly with materials, exact structures, and solid-liquid interfacial properties. Peptoids, or N-substituted glycines, with highly tunable sequence, and consequently functionality, contributes to a rising group of promising structure-directing chemical additives. We find that an amphiphilic peptoid Nce3Ncp6, composed of three carboxyl side chains and six chlorobenzyl side chains, can effectively assist the formation of Au nanostars in water. These nanostars adopt an unconventional five-fold twinned structure, comparing to a classical decahedron, as the twin planes extend to the grooves of the star, instead of the vertices on a decahedron. Since these nanostars are majorly {111}-faceted, it is hypothesized that these shapes are formed as a result of a preferred binding to Au(111) facets from peptoid Nce3Ncp6.

We use all-atom molecular dynamics (MD) simulations, together with enhanced sampling methods, such as Parallel Bias metadynamics (PBMetaD) and umbrella sampling, to study the adsorption of peptoid Nce3Ncp6 at the Au-water interface and unravel how surface adsorption can affect shape revolution. Using PBMetaD, we find that Nce3Ncp6 strongly prefers Au(111) over Au(100), and the origin of this binding preference can be dissected into two aspects. First, while water forms layers near both surfaces, the packing of the first-layer water molecules is much more ordered on Au(100) than Au(111), which protects the Au(100) facet from molecular adsorptions. Second, aromatic rings significantly favor Au(111) surface due to surface electron density distribution, and their energetic preference is the major contribution to the energetic selectivity of Nce3Ncp6. The carboxyl side chains on the peptoid, however, bind weakly and non-selectively to either facet. As a result, the adsorption of Nce3Ncp6 can largely passivate the Au(111) facet thermodynamically, counterbalance the energy associated with the twin and stabilize the five-fold twinned structure. Due to the amphiphilic and facet-selectivity nature of the peptoid, its packing on surfaces can influence the deposition of solution-phase atoms towards different facets. Using umbrella sampling, we investigated the steric effect from the peptoids adsorbed on Au(100) and Au(111). Weak adsorption results in loosely packed peptoids on Au(100), allowing a higher deposition flux towards Au(100), and thus grows the vertices on the nanostars. In sum, the formation of five-fold Au nanostar is regulated both thermodynamically and kinetically by the strongly selective binding of peptoid Nce3Ncp6 on Au surfaces.

611793 Tlmutation: Predicting the Effects of Mutations Using Transfer Learning.

Abstract:

A reoccurring challenge in bioinformatics is predicting the phenotypic consequence of amino acid variation in proteins. With the recent advancements in sequencing techniques, sufficient genomic data has become available to train models that predict the evolutionary statistical energies, but there is still inadequate experimental data to directly predict functional effects. One approach to overcome this data scarcity is to apply transfer learning and train more models with available datasets. In this talk, we present a set of transfer learning algorithms we call TLmutation, which implements a supervised transfer learning algorithm that transfers knowledge from survival data of a protein to a particular function of that protein. This is followed by an unsupervised transfer learning algorithm that extends the knowledge to a homologous protein. We explore the application of our algorithms in three cases. First, we test the supervised transfer on 17 previously published deep mutagenesis datasets to complete and refine missing data points. We further investigate these datasets to identify which mutations build better predictors of variant functions. In the second case, we apply the algorithm to predict higher-order mutations solely from single point mutagenesis data. Finally, we perform the unsupervised transfer learning algorithm to predict mutational effects of homologous proteins from experimental datasets. These algorithms are generalized to transfer knowledge between Markov random field models. We show the benefit of our transfer learning algorithms to utilize informative deep mutational data and provide new insights into protein variant functions. As these algorithms are generalized to transfer knowledge between Markov random field models, we expect these algorithms to be applicable to other disciplines.

599372 Shifting Paradigms: Using Molecular Dynamics with Machine Learning for the Calculation of Thermodynamic Properties of Fluids

Abstract:

The accurate prediction of fluid properties holds an essential role for chemical engineering, with ubiquitous applications in many areas from process design, reaction engineering to fluid dynamics. For this purpose, the current chemical engineering folklore employs empirical and/or theory-based correlations based on a very reduced number of experimental observations. The limitations are obvious, particularly when extrapolating to unknown compounds and/or thermodynamic conditions remote from those represented by the experiments. We propose a new workflow, where we strive to combine molecular simulations (validated against experiments) to deliver an extensive and reasonably accurate database of thermodynamic properties with machine learning (ML) algorithms to provide for the synthesis and correlation of the information. We suggest that this new paradigm can, in some instances, take over the classical experiment/equation of state tools currently in use in the field. As a proof-of-concept we explore the use of the SAFT- Mie force field[1]to generate critical properties, vapor pressures and saturated densities of a representative family of common fluids and correlate them with ML techniques.

In this work, we employ machine learning algorithms such as artificial neural networks (ANN) and Gaussian process regression (GPR). The pros and cons of the models created by each technique are analysed, considering the amount of data required, the accuracy of the model and other further applications and drawbacks of each technique. With prior knowledge of the underlying theory relating these thermodynamic properties, hybrid models are created in the sense of using suitable data transformations with theoretical basis.

Using the SAFT- Mie coarse-grained force field, molecules can be described as chains of identical spherical segments employing four key molecular descriptors . A key aspect of the model is that there is an existing equation of state (EoS)[2]that can reproduce the outputs of the molecular simulations with quantitative accuracy. By generating pseudo-experimental data for a range of these parameters, as a proof-of-concept, we attempt to replicate the SAFT-VR Mie equation of state with machine learning, to predict critical properties , vapour-liquid phase equilibrium properties and supercritical densities from these molecular descriptors and state properties.

The comparison of PvT properties between the EoS and the ML model provides a benchmark to the viability of different machine learned models in predicting fluid phase properties. Different properties present different complexity of machine-learned model, with critical properties taking zero state properties as input being the simplest model, while supercritical properties requiring two state properties (temperature and pressure). While both ANN and GPR models are able to correlate with high degree of statistical performance with R^2values above 0.99, the selection of activation functions and data transformation methods is essential. For a vapour pressure machine-learned model, the direct correlation between and results in a model that appears to give good R^2score, but corresponds to extremely large average absolute deviation (AAD), as pressure varies across different orders of magnitude. Alternatively, although the use of the Clausius-Clapeyron relationship and training a model to correlate and might appear worse near the critical point, the overall performance in AAD improves significantly. Similarly, selecting the correct activation function improves the curves of the

VLE envelope. Looking at individual saturated density curves, models with less smooth activation functions (such as the Matern kernel for GPR, or the ELU function for ANN) presents irregular shapes despite giving a statistically acceptable performance. This is because the machine is "informed" of the general shape of the model through the activation functions, which is essential to produce a good correlation. We show how the ideal activation function in this case for ANN would be the *tanh*activation function, while similarly the radial basis function is the ideal kernel for GPR.

A pathway towards the correlation of fluid properties from molecular dynamics simulated data to a machine-learned model is being drawn out, with applications from both machine learning techniques. Comparing the two machine learning algorithms used, GPR requires much less data to achieve quantitative accuracy, and it is a more sophisticated technique as it calculates the variance of the predicted model which is useful for error analysis. However, it becomes increasingly expensive, from a computational point of view, as the complexity of the problem escalates with more input dimensions. In this sense, ANN is a more robust and flexible technique but requires a much larger set of data to be generated. Discussion along these lines is included in the presentation.

References:

[1] Müller, E. A.; Jackson, G. (2014). Force-field parameters from the SAFT-equation of state for use in coarse-grained molecular simulations. *Annu. Rev. Chem. Biomolec. Eng.*, 5(1), 405–427.

[2] Lafitte, T., Apostolakou, A., Avendaño, C., Galindo, A., Adjiman, C.S., Müller, E.A., Jackson, G. (2013). Accurate statistical associating fluid theory for chain molecules formed from Mie segments. *J Chem. Phys.*, 139(15), 154504.

599712 Combining Molecular Dynamics and Machine Learning to Predict Interfacial Properties

Abstract:

Molecular dynamics (MD) simulations are a powerful tool for analyzing the properties and behavior of materials in solvent environments by explicitly representing atomistic degrees of freedom and accounting for intermolecular interactions. However, MD simulations can require lengthy runtimes that limit the high-throughput screening of system properties. Moreover, the analysis of MD simulation output typically requires the implementation of techniques that rely on human insight to relate the properties of a statistical ensemble to relevant macroscopic properties. Alternatively, machine learning has emerged as a valuable tool to interpret molecular simulation data. Here, we present computational techniques that use machine learning to analyze molecular simulations of complex solvent environments. We focus on quantifying the hydrophobicity of functionalized interfaces by utilizing deep learning to analyze MD-derived interfacial water configurations. By selecting an appropriate representation of the MD output data, we demonstrate that a 2-dimensional convolutional neural network (2D CNN) can infer spatial correlations between solvent molecules and map these features to thermodynamic observables. We systematically vary the representation of the MD input data to determine the impact on the accuracy of the ML predictions. Once trained, we find that the 2D CNN can predict dewetting free energies with minimal simulation input, facilitating the rapid screening of interfacial hydrophobicity for a range of surface compositions. We further compare the accuracy of alternative ML architectures, including 3D CNNs that incorporate temporal correlations in our deep learning framework and graph neural networks that incorporate the network connectivity of water hydrogen bonds. Together, these methodologies demonstrate that ML analysis of spatial and temporal correlations that are encoded within the solvent environment can rapidly predict interfacial hydrophobicity to guide materials design.

609813 Data Science Based Electrolyte Design for Organic Redox Flow Batteries

Abstract:

Ability to design materials for targeted functionality rather than traditional intuitive and trial-by-error based methods can bring enormous societal and technological progress. In this context, exploring the large space of potential materials for energy storage device is computationally intractable. In particular, the discovery of organic redox active species - which dictates the energy density, specific capacity and cycle life is critical step in realizing viable redox flow battery (RFB) technology - will be taunting challenge considering billions of organic structural possibilities. Traditional methodology relies on combinatorial approach of additives, functional groups and solvent composition to achieve higher solubility and stability of the specific redox molecule. However, considering the wide range of materials selection, it is imperative to build an artificial intelligence-based design formulation that will help us bottom up design of RFB electrolytes. In recent years, our team has been building database and scientific framework for electrolyte design formulations based on chemistry informed (e.g. solvate structure and inherent properties) machine learning approach. This framework helps us in solvent and functional group selection process, and ultimately enhance the solubility and stability through inverse molecular design. We will discuss our efforts and achievements related to the data science-based framework of electrolyte design formulations applicable for organic molecule based redox flow battery systems.
Ability to design materials for targeted functionality rather than traditional intuitive and trial-by-error based methods can bring enormous societal and technological progress. In this context, exploring the large space of potential materials for energy storage device is computationally intractable. In particular, the discovery of organic redox active species - which dictates the energy density, specific capacity and cycle life is critical step in realizing viable redox flow battery (RFB) technology - will be taunting challenge considering billions of organic structural possibilities. Traditional methodology relies on combinatorial approach of additives, functional groups and solvent composition to achieve higher solubility and stability of the specific redox molecule. However, considering the wide range of materials selection, it is imperative to build an artificial intelligence-based design formulation that will help us bottom up design of RFB electrolytes. In recent years, our team has been building database and scientific framework for electrolyte design formulations based on chemistry informed (e.g. solvate structure and inherent properties) machine learning approach. This framework helps us in solvent and functional group selection process, and ultimately enhance the solubility and stability through inverse molecular design. We will discuss our efforts and recent results related to the data science-based framework of electrolyte design formulations applicable for organic molecule based redox flow battery systems.

610519 Development of an Artificial Neural Network Model to Predict Electrical Conductivity of Pure Ionic Liquids and Ionic Liquid-Ionic Liquid Mixtures

Abstract:

Ionic liquids are room temperature ionic solvents characterized by low volatility, designer tunability, high thermal and electrochemical stability. These properties make ionic liquids ideal candidates for the replacement of organic solvents as electrolytes in Li-ion batteries. Despite their favorable attributes, high viscosity of many ionic liquids is a hindrance to exploiting their full potential for battery applications. One way to tackle this challenge is to develop novel ionic liquids by altering the identity of the cation, anion, or substituents on the ions. The enormous chemical space for such a manipulation is extremely difficult, if not impossible, to navigate efficiently via experimental or molecular simulation-based approaches. In such instances, machine learning techniques offer an exciting alternative, if sufficient data is available to train a model.

In this presentation, we will describe our efforts at developing a machine learning model based on the ionic conductivity data obtained from the NIST ILThermo Database. A feed forward artificial neural network (FFAAN) comprised of an input layer, a hidden layer and an output layer is obtained from the data on ~100 ionic liquids. We will demonstrate that the FFAAN is able to capture the data with a high degree of accuracy over five orders of magnitude. The machine learning model is then used to predict the ionic conductivity data for ~1200 pure ionic liquids obtained by combining the unique cations and anions in the database. We will also show how the machine learning model can be employed to predict the ionic conductivity for ~840,000 ionic liquids. The validity of such predictions will be demonstrated by comparing the ionic conductivity of those binary ionic liquids for which experimental data is available. Furthermore, we will report on 5000 potential binary ionic liquid-ionic liquid mixtures for which there exists a possibility for obtaining ionic conductivity higher than those of the pure ionic liquids making up the binary ionic liquid mixtures.

598191 Incorporating Long-Range Interactions in Machine Learning Models of Water and Aqueous Electrolyte Solutions

Abstract:

In recent years, atomistic machine learning models have become increasingly popular in molecular simulations, given their ability to combine the accuracy of quantum mechanical representations with the speed and efficiency of classical potentials. These models are capable of learning highly complex and multi-dimensional interactions within a local environment but face challenges in capturing long-range behavior. In this work, we train deep neural networks to interatomic Potential Energy Surfaces to construct many-body classical potentials that accurately represent the structure and dynamics of water and electrolyte systems. We introduce a formalism to represent long-range electrostatics by decomposition of the Coulombic Ewald energy from the local environment representation. We demonstrate the effectiveness of this approach by predicting electrolyte solution properties and vapor-liquid coexistence behavior, which highlights the versatility of our models in representing heterogeneous and charged systems.

600934 Fully Convolutional Neural Network Models for Materials Science Applications

Abstract:

In silico design can be an efficient and cost effective method for discovering new materials. Traditionally, quantum mechanics, classical approaches, and quantitative structure property relationships (QSAR) are used to design materials using the structure of the materials as an input. Although these methods are helpful in designing materials, they all suffer from a common drawback: they estimate properties from an input structure which usually necessitates iteration between synthesis and modeling to achieve the desired performance metric. Moreover, that process is also dependent on the ingenuity of design team to identify potential materials for computational evaluation which may be very different structurally from the benchmark material. Here we describe a generative deep learning approach to address some of these drawbacks: a fully convolutional neural network (FCNN) molecular generative model. The efficiency and reconstruction accuracy of this method was found to be comparable with Junction Tree variational autoencoder (JT-VAE) and superior to a SMILES-based variational autoencoder (CVAE). Furthermore, this approach (p-FCNN) is tuned to generate molecules which incorporate a larger fraction of heteroatoms (heavy atoms other than C's) compared to those from CVAE. Detailed analysis will be provided to understand the relative distribution of the newly generated molecules vs. the training set. In addition, the latent space generated from FC-NN will be organized according to ALogP and the response surface will be smoothed using a Gaussian process model. Gradient search will be performed on the organized latent space to extract molecules near the optimal value of ALogP. In conclusion, FCNN and p-FCNN can be considered as an efficient generative deep learning approach to discover materials with an optimized performance metric.

599439 Deep Learning for Characterizing and Modeling the Dynamics of Complex Colloidal Self-Assembly Systems

Abstract:

Self-assembly (SA) is the process by which discrete components (e.g., nanoparticles in solution) spontaneously organize into an ordered structure [1]. The spontaneous self-organization central to SA enables "bottom-up" materials synthesis, which allows for manufacturing advanced, highly ordered crystalline structures in an inherently parallelizable and cost-effective manner. Thus, SA can create new avenues for highly scalable, economical manufacturing of novel metamaterials with unique optical, electrical, or mechanical properties [2-4].

SA is an inherently stochastic (i.e., random) process prone to kinetic arrest [2]. This leads to variability in materials manufacturing and possibly high defect rates, which can severely compromise the viability of using SA to reproducibly manufacture advanced materials. However, creating a systematic methodology to characterize and understand the highly nonlinear and stochastic SA dynamics responsible for kinetic arrest remains an open grand challenge. Previous attempts to characterize kinetically arrested states have either relied on physics-based order parameters that cannot accurately resolve structures in the face of thermal fluctuations or unsupervised learning techniques based on diffusion maps that provide incomplete physical interpretations of the underlying kinetics [5-7]. Moreover, all attempts to model SA dynamics either involve physics-based strategies based on the chemical master equation or parameter estimation strategies based on Bayesian Inference, neither of which are computationally tractable for practically sized SA systems [7-8]. Due to their strong adaptability and generalization properties in learning, strong noise tolerance, and ability to solve exceptionally large problems, deep neural networks (DNNs) are ideal candidates for investigating the complex, stochastic, nonlinear dynamics of SA systems [9].

In this work, we develop a type of DNN called a variational autoencoder [10] to translate the positions of individual SA particles into an easily interpretable low-dimensional space. The autoencoder provides an explicit mapping between the particle positions and the low-dimensional space. We extrapolate the knowledge from this mapping to understand the relevant symmetries that distinguish the different environments possibly present in the system and determine a rigorous physical interpretation of the low-dimensional space. We calculate low-dimensional free energy landscapes from SA trajectory data and use the relative free energies to identify the kinetically arrested states into which the system may assemble. We leverage a second DNN to model the evolution of the SA over time and predict the likelihood that SA systems avoid kinetic traps or reach target states under certain external conditions (e.g., temperature). The main contribution of this work is the development of a systematic framework to not only fully characterize kinetic traps, but also analyze exactly how these kinetic traps can be avoided in order to reproducibly reach high-value structures.

We demonstrate the efficacy of the proposed framework on an *in-silico* three-dimensional system of self-assembling DNA functionalized colloids that have shown enormous promise for sensing and photonics applications [11,12]. The system is especially prone to kinetic arrest due to the complexity of its underlying (and often competing) free energy surface, which includes repulsive interactions among the underlying silica particles, repulsive interactions due to ssDNA chain overlap, and attractive interactions due to ssDNA hybridization. We characterize and analyze the relative importance of these kinetic traps and demonstrate how changes in temperature (which turn ssDNA bonds "on" or "off") can be used to avoid kinetic traps and reach target structures.

References

[1] Whitesides, G. M., & Grzybowski, B. (2002). Self-assembly at all scales. *Science*, *295*(5564), 2418-2421.

[2] Paulson, J. A., Mesbah, A., Zhu, X., Molaro, M. C., & Braatz, R. D. (2015). Control of self-assembly in micro-and nano-scale systems. *Journal of Process Control*, *27*, 38-49.

[3] Grzelczak, M., Vermant, J., Furst, E. M., & Liz-Marzán, L. M. (2010). Directed self-assembly of nanoparticles. *ACS nano*, *4*(7), 3591-3605.

[4] Liddle, J. A., & Gallatin, G. M. (2016). Nanomanufacturing: a perspective. *ACS nano*, *10*(3), 2995-3014.

[5] Long, A. W., & Ferguson, A. L. (2019). Landmark diffusion maps (L-dMaps): Accelerated manifold learning out-of-sample extension. *Applied and Computational Harmonic Analysis*, *47*(1), 190-211.

[6] Reinhart, W. F., & Panagiotopoulos, A. Z. (2018). Automated crystal characterization with a fast neighborhood graph analysis method. *Soft matter*, *14*(29), 6083-6089.

[7] Tang, X., Rupp, B., Yang, Y., Edwards, T. D., Grover, M. A., & Bevan, M. A. (2016). Optimal feedback-controlled assembly of perfect crystals. *ACS nano*, *10*(7), 6791-6798.

[8] Lakerveld, R., Stephanopoulos, G., & Barton, P. I. (2012). A master-equation approach to simulate kinetic traps during directed self-assembly. *The Journal of chemical physics*, *136*(18), 184109.

[9] Jain, A. K., Mao, J., & Mohiuddin, K. M. (1996). Artificial neural networks: A tutorial. *Computer*, *29*(3), 31-44.

[10] Ohno, H. (2019). Auto-encoder-based generative models for data augmentation on regression problems. *Soft Computing*, 1-11.

[11] Pretti, E., Zerze, H., Song, M., Ding, Y., Mahynski, N. A., Hatch, H. W., ... & Mittal, J. (2018). Assembly of three-dimensional binary superlattices from multi-flavored particles. *Soft matter*, *14*(30), 6303-6312.

[12] Pretti, E., Mao, R., & Mittal, J. (2019). Modelling and simulation of DNA-mediated self-assembly for superlattice design. *Molecular Simulation*, *45*(14-15), 1203-1210.

606830 AI Based Analysis for Graphene Synthesis - from Size Measurement to SEM Image Generation

Abstract:

Graphene is a single atomic layer compound made up of pi (π) bonded carbon atoms. It is an excellent material used in the production of electronic and photonic components such as flexible transparent electrodes and field-effect transistors, due to its high mechanical strength, electrical conductivity, and mobility [1]. In this study, to synthesize high-quality graphene, a chemical vapor deposition experiment was conducted by using CH_4 as a precursor on Cu substrate at high temperatures [2]. The size of graphene, coverage, domain density and aspect ratio, which vary depending on the synthesis conditions including temperature, annealing time, growth time, and hydrogen supply were measured and analyzed. Owing to the variations in the shapes of graphene grains, a manual measuring technique is often used to calculate the size. Not only is this method laborious but also inefficient and resource intensive. As such, a faster, efficient, inexpensive, and standard measuring method was developed by employing Region-proposal Convolutional Neural Network(R-CNN): an objective detection algorithm [3]. In this strategy, the graphene grains were assumed to be hexagonally shaped. Consequently, the aspect ratio was also estimated by the model. Monochrome processed graphene imprinted SEM images were used as inputs by k-means clustering for the R-CNN training [4]. The R-CNN predicted size and aspect ratio were in the error margin of about 0-10% when validated against experimental data. Subsequently, to ascertain the relationship between the process variables and the size, coverage, and domain density, an Artificial Neural Network (ANN) algorithm was modelled [5]. A Support Vector Machine (SVM) was further developed to investigate the effect of the process conditions on graphene shapes using the aspect ratio as target variable [6]. Finally, by developing a Generative Adversarial Network (GAN) [7], hexagonally shaped graphene grains were generated as black and white monochrome images with the same domain density as in the experimental data. A complete network of all the developed algorithms (R-CNN, ANN, SVM, GAN) was created to generate the graphene images. To obtain the same image color and outlook just as in the experimental SEM images, a Pix2Pix model was developed to transform the monochrome images into colored real SEM images [8]. By comparison, the ANN predicted images showed high agreement with the experimentally generated ones. As a result, we were able to find optimum experimental condition efficiently using this newly developed model, having a deeper insight into graphene CVD synthesis and predicting results in the form of images.

Literature cited:

[1] Phaedon Avouris and Fengniam Xia, “Graphene applications in electronics and photonics” ,MRS Bulletin,vol.37,pp. 1225-1234, 2012.

[2] Xuesong Li, Wei Wei Cai, Jinho An, Dongxing Yang,Richard Piner, Aruna Veamakanni, Inhwa Jung, Rodney S.Ruoff “Large-Area Synthesis of High Quality and Uniform Graphene Films on Copper Foils”, *Science*, vol. 324, pp. 1312-1314, 2009.

[3] Ross Girshick, Jeff Donahue, Trevor Darrel, Jitendra Malik “Rich Feature Hierarchies for Accurate Object Detection and Semantic Segmentation”, The IEEE Conference on Computer Vision and Pattern Recognition (CVPR), pp. 580-587, 2014

[4] Ray Siddheswar, Turi R.H. “Determination of number of clusters in k-means clustering and

application in colour image segmentation", Proceedings of the 4th international conference on advances in pattern recognition and digital techniques, pp.137-143, 1999

[5] Kenji Suzuki, "Artificial Neural Networks - Methodological Advances and Biomedical Applications", ISBN 978-953-307-243-2, pp. 2-14, 2011

[6] Hsu, C.-W., Chang, C.-C., and Lin, C.-J. "A practical guide to support vector classification. Tech. rep., Department of Computer Science", National Taiwan University, 2003

[7] Goodfellow, Ian J, Pouget-Abadie "Generative adversarial nets", Advances in Neural InformatiProcessing Systems, pp.2672-2680, 2014

[8] P. sola, Phillipzhu, Jun Yan Zhou "mage-to-Image Translation with Conditional Adversarial Networks", Proceedings - 30th IEEE Conference on Computer Vision and Pattern Recognition, pp. 5967-5976, 2017

602682 Infrared Spectra Prediction with Machine Learning

Abstract:

Spectroscopy is an important tool for manufacturing for ascertaining purity in biological systems and drug manufacturing. Prediction of unknown spectra would be a powerful tool for extending existing spectral identification databases and evaluating novel molecules for which there exist no standards. Machine learning algorithms using deep neural networks have been used for the interpretation of spectra in microbiology and food production for the last ten years. The prediction of the spectra themselves is a more recent area of research, with new efforts to predict IR spectra, proton NMR spectra, and mass spectra. It is the purpose of this present effort to use machine learning to predict IR spectra for novel molecules, improving on previous such work by including effects beyond calculated harmonic frequencies and using learned molecular fingerprints.

A model for IR spectral prediction has been developed, requiring only the input of a molecular graph structure provided in the form of a SMILES code. The chemical representation is processed through a message passing neural network followed by a feed forward neural network for prediction of IR spectra, an adaptation of the Chemprop code for molecular property prediction. The model is pre-trained using quantum chemistry calculations for molecules sampled from the PubChem database to learn the molecular harmonic vibration modes. During pretraining, active learning is used to explore the chemical space more efficiently and prioritize molecules which will most improve the model. Further training is performed using experimentally collected spectra available in open-access databases from the National Institute of Standards and Technology (NIST) and the National Institute of Advanced Industrial Science and Technology (AIST). The model allows for predictions of spectra in the gas phase and in supported condensed phases.

607833 Optimization of Thermal Conductivity at Interfaces Using Learning Algorithms

Abstract:

In material science, we are frequently interested in understanding the properties and design implications of material at interfaces. These interfaces can be manipulated to improve the desired characteristics of the bulk material. Optimization of thermal transport across the interface of two different materials is critical to micro/nanoscale electronic, photonic, and phononic devices. However, interfacial thermal conductivity optimization is a high-dimensional problem due to the number of factors affecting it. Non-equilibrium molecular dynamics (NEMD) has been used to investigate the thermal transport across a solid-solid interface between different materials. Despite NEMD being an effective approach to this problem, scanning the large parameter space is unrealistic due to high computational cost. Alternatively, machine learning is a cost-effective and time-efficient method that can investigate this interfacial thermal transport phenomenon.

In this talk, we describe our efforts to understand and optimize the impact of interfacial atomic defects and intermixing on the thermal transport across an Al/Si junction. We developed a learning-based framework to optimize over a parameter space that would be prohibitively large in the brute force approach. Using this technique allows us to accumulate knowledge of the system of a given type of atoms and store this information into a neural network. A scalable framework was built to allow utilization of the available high-performance computing resources, thereby accelerating the learning by directly interfacing with the LAMMPS molecular dynamics simulation package. We found that mixing 3-4 monolayer of atoms near the interface can significantly increase the thermal conductance. However, when inter atomic mixing layers get thicker, it induces a negative effect on the thermal conductance due to phonon scattering. The model is able to find the optimal mixing length and mixing fraction relatively fast. This work can be extended to more systems by inclusion of descriptors and a high-fidelity surrogate model.

606146 High-Throughput Screening of Small-Molecule Libraries in Search of New Antibiotics

Abstract:

In the face of increasing bacterial resistance to all known antibiotics, efforts to discover new classes of antibiotics have languished, creating an urgent need to accelerate the antibiotic discovery pipeline. A major deterrent in the discovery of new antibiotics is the limited permeability of molecules across the bacterial envelope. Notably, the Gram-negative bacteria have nutrient specific protein channels (or porins) that restrict permeability of non-essential molecules including antibiotics. To accelerate the antibiotic discovery process, we have developed the ComputationaL Antibiotic Screening Platform (CLASP) for high-throughput screening of potential drug molecules through the porins. The CLASP takes advantage of coarse grain (CG) resolution, advanced sampling techniques, and a parallel computing environment to maximize its performance. The CLASP yields comprehensive thermodynamic and kinetic output data of a potential drug molecule within a few hours of wallclock time. Its output includes the potential of mean force profile, energy barrier, rate constant, half-life of the molecule's transport process, contact analysis of the molecule with the pore-lining residues, and the orientational analysis of the molecule in the porin channel. In our first CLASP application, we report the transport properties of six carbapenem antibiotics—biapenem, doripenem, ertapenem, imipenem, meropenem, and panipenem—through OccD3, a major channel for carbapenem uptake in *P. aeruginosa*. The OccD3 channel is a nanopore with a constriction zone lined by a combination of polar and bulky

nonpolar residues. These residues provide size selectivity to the porin and make small molecule uptake in *P. aeruginosa* cells much harder. The CLASP is designed to screen small molecule libraries with a fast turnaround time to yield structure-property relationships to discover antibiotics with high permeability. The CLASP will be freely distributed to enable accelerated antibiotic drug discovery.

608430 Binding Free Energy Calculation on Human Beta Defensin Type 1 (hBD-1) and Type 3 (hBD-3) on Bacterial Membrane

Abstract:

Human beta defensins (hBD) are small antimicrobial peptides acting as an important part of human innate immunity. They have antimicrobial activities against different microorganisms including bacteria, virus, fungi. Several human beta defensins have been isolated so far and all of them have the ability to kill or inhibit bacteria at low salt concentrations. It is believed that hBDs interacts with bacteria by disrupting its membrane. So, it is important to investigate its interaction with the bacterial membrane in molecular details to understand the antimicrobial mechanism. Among the hBDs, hBD-3 is the most potent and found to be active against both Gram-positive and Gram-negative bacteria even at high salt concentrations. It has 45 residues and a charge of +11. hBD-1 has 36 residues and a charge density of +7. hBD-1 shows antimicrobial activity against some gram-negative bacteria at low salt concentration, but relatively less potent against gram-positive bacteria. In this study, we are going to investigate the binding free energy of the residues on hBD-3 and hBD-1, with a model bacterial membrane consisting of pure POPG lipids; because both Gram-positive and Gram-negative bacterial membranes consist of POPG lipids at high concentrations, 67% and 33% of the total phospholipids respectively. Comparing the binding free energy results of hBD-3 with hBD-1 might give an insight of the residues which make hBD-3 potent against both gram-positive and gram-negative bacteria. As it is quite complicated to calculate the absolute binding free energy of the residue with membrane, the relative binding free energy of protein will be calculated resulting from the point mutation of each residue to Alanine. A thermodynamic cycle has been designed to calculate the difference in the binding free energy. The alchemical free energy changes will be calculated by running two free energy perturbation simulations of the mutation and the difference between the mutation free energy gives the difference in the binding free energy of the protein in wildtype and the mutant. The simulations have been performed using NAMD program at a temperature of 300.15 K and a pressure of 1 atm, dividing the whole process into 20 different thermodynamic windows and running each window for 3ns in the bound state simulation and 2ns for unbound state simulation. Based on the free energy results of hBD-3, it is found that Arginine 17, Arginine 36 and Arginine 43 bind quite strongly with the POPG lipid compared to the other residues contributing to a total binding free energy of – 9.1297 kcal/mol out of -28.388 kcal/mol. The work emphasized the contribution of positively charged Arginine residues to the binding between hBD-3 and bacterial membrane.

606356 Quantum-Based Modeling of Radiation Damage in Siloxane Polymers

Abstract:

Exposing materials to ionizing radiation is a reliable means of sterilization and is a common approach taken in accelerated aging experiments. Initial atomistic-level radiation damage in chemically reactive materials such as silicones is thought to induce a sequence of network-altering events that lead to undesirable macroscale degradation, including permanent set and mechanical failure. We develop a multiscale approach based on semiempirical quantum molecular dynamics (QMD) to predict and analyze primary radiation damage in polydimethylsiloxane (PDMS) and gauge the macroscale consequences of that damage by connecting with a constitutive model. Large ensembles of QMD simulations are used to predict the initial reaction cascades that follow from primary knock-on atom radiation events. A graph-based analysis is developed to automatically identify changes to the backbone structure and quantify mechanically relevant network alterations including formation of cross-links and chain scissions. Distinguishing characteristics of radiation coupling to different parts of the PDMS backbone are explored. Results from QMD modeling and the graph-based analysis are used to inform a simplified constitutive model for the mechanical response of irradiated elastomers. This work was performed under the auspices of the U.S. Department of Energy by Lawrence Livermore National Laboratory under Contract DE-AC52-07NA27344. Approved for unlimited release, LLNL-ABS-809007.

601194 Understanding V2+/V3+ Complexation and Reaction on Glassy Carbon in Acidic Electrolytes from First Principles

Abstract:

The rapid rise in global energy demand and desire for lower carbon emissions has prompted the use of renewable sources for electricity generation. With intermittent solar and wind power expected to generate 50% of all electricity by 2050 [1], energy storage solutions are needed to match energy supply with consumer demands. All-Vanadium Redox Flow Batteries (VRFBs) are a promising technology for grid energy storage because of their scalability, long lifetimes, and low maintenance. Despite these advantages, VRFBs are still limited by high costs, which can be lowered by increasing the current densities of the kinetically limiting V^{2+}/V^{3+} redox couple. However, there is still a lack of fundamental understanding of the V^{2+}/V^{3+} reaction mechanism, particularly the role of electrolytes in the charge transfer. In this talk, we elucidate the structure of vanadium ions in common acidic electrolytes and probe their redox kinetics at carbon electrode surfaces using Density Functional Theory (DFT) modeling and comparing with experimental measurements. We model vanadium complexes at the electrode surface to gain insight into the processes occurring at the atomic level and to help interpret experimental observations.

In our recently published work [2], we elucidate vanadium complexation with anions in common HCl and H_2SO_4 electrolytes by comparing experimental UV-Vis spectra to Time-Dependent Density Functional Theory (TDDFT) calculations, as well as DFT-predicted free energies of V^{3+}-anion complexes in solution. DFT predictions using the hybrid B3LYP-D3 functional and an explicit/implicit solvent model indicate that the free energies of the complexes follow the order of $[V(H_2O)_5SO_4]^+ < [V(H_2O)_4Cl_2]^+ < [V(H_2O)_6]^{3+}$, in agreement with our experimental observations. The V^{2+}/V^{3+} kinetics are measured to be ~2.5 times faster in HCl than in H_2SO_4 or HCl/H_2SO_4, attributed to the presence of $[V(H_2O)_4Cl_2]^+$, where adsorbed Cl (*Cl) acts as a bridge between the carbon electrode and the vanadium ion. Previously, similar enhancement has been reported for Cr^{2+}/Cr^{3+} redox couple [3], attributed to the polarizability of the bridge. A bridging mechanism through *Cl is supported by even faster redox kinetics in HBr than HCl, where adsorbed *Br being a better bridge than *Cl enhances the charge transfer.

To understand the role of halide bridging ligands on the charge transfer kinetics of the V^{2+}/V^{3+} reaction, we computed adsorption energies of halides (Cl, Br, and I) on a glassy carbon surface as a function of coverage using the PBE-D3 functional. These adsorption energies were used as inputs to a generalized computational hydrogen electrode model [4], where we determined that halide coverages on glassy carbon follow the trend $\theta_I > \theta_{Br} > \theta_{Cl}$ at the V^{2+}/V^{3+} reduction potential (E^0 = -0.255 V). We also compare the energies of vanadium complexes adsorbed to a glassy carbon surface through OH, Cl, Br, and I ligands, representing the proposed reaction intermediates in H_2SO_4, HCl, HBr, and HI, respectively (Figure 1). The decreasing adsorption strength of $[*Cl\text{-}V(H_2O)_5]^{2+} > [*Br\text{-}V(H_2O)_5]^{2+} > [*I\text{-}V(H_2O)_5]^{2+}$ supports our hypothesis that V^{2+}/V^{3+} redox kinetics is dependent on the stability of the adsorbed intermediate. These findings open avenues in electrolyte engineering to increase V^{2+}/V^{3+} redox kinetics in VRFBs.

References:

[1] Bloomberg New Energy Outlook, **2019** <https://bnef.turtl.co/story/neo2019/>

[2] Agarwal, H., Florian, J., Goldsmith, B., Singh, N. (2019). *ACS Energy Letters*, **2019**, 4, 2368–2377

[3] Sykes, A. G. Further Advances in the Study of Mechanisms of Redox Reactions. *Adv. Inorg. Chem. Radiochem.* **1968**, 10, 153–245.

[4] Gossenberger, F. et. al., **2015**. *Surface Science*, 631, 17-22

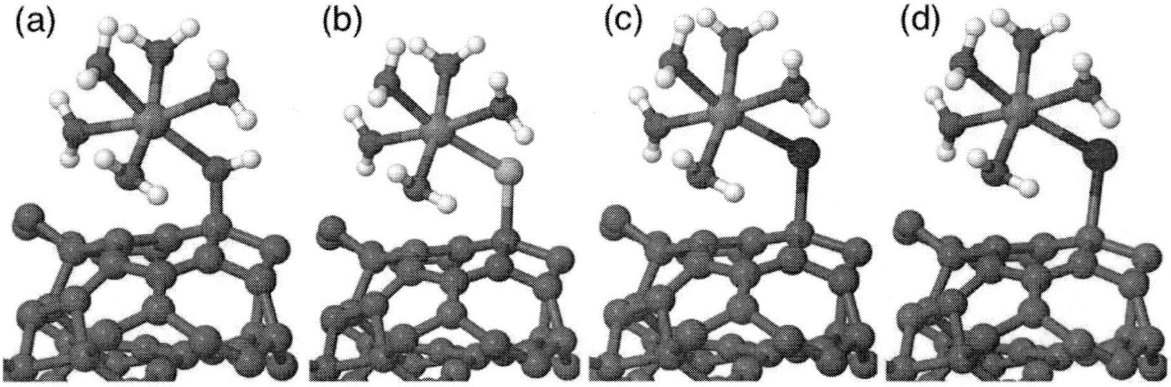

Figure 1. Proposed V^{2+}/V^{3+} reaction intermediates on glassy carbon in (a) H_2SO_4, (b) HCl, (c) HBr, and (d) HI electrolytes. Atom color legend: Dark gray = C, Red = O, White = H, Light gray = V, Green = Cl, Magenta = Br, and Purple = I.

603759 A Molecular Model for Reactive Fouling in Steam-Cracker Separation Trains

Abstract:

Polymeric fouling is a ubiquitous problem in the separation train downstream of steam crackers used to pyrolyze ethane and other saturated feeds into alkenes (olefins). Macromolecules form and deposit in downstream equipment, forming films on those surfaces over time, resulting in sub-optimal operation and eventually forcing plant shut down. A variety of empirical approaches have been developed and used, with mixed success, to ameliorate this problem. However, there is limited fundamental understanding on the origin of fouling. Because the root cause of fouling is often not known, it is difficult to determine the best mitigation strategy.

In this work, we lay the foundations for understanding polymer fouling with a multiscale multiphase model for the deposit growth process. First the composition at different locations in the separation train is computed using ASPEN. Then a reaction network suitable for modeling the kinetics occurring in that chemical mixture is constructed using the Reaction Mechanism Generator (RMG), a software package that constructs detailed chemical reaction mechanisms by a rate-based model enlarging algorithm. The rates of key reactions and the thermochemistry of critical reaction intermediates are computed using high level quantum chemistry methods; in the present case these reactions include Diels-Alder, ene, and free radical reactions. This extensive, detailed list of elementary reactions is embedded within a multiphase and multiscale deposition model, represented as a set of equations consisting of foundational balances and constitutive relations to describe mesoscale-continuum dynamics. Simulations are conducted with Julia, a high-performance scientific computing language. Computed sensitivities of the simulated deposit growth rates to the model parameters are used to direct iterative refinement of the model.

The model presented here focuses on diene-driven deposit growth at specific reaction conditions. But the general modeling framework presented here, which combines classical analyses of phase equilibria, transport, and material properties with automated reaction mechanism generation and quantum chemistry calculations of reaction kinetics, is expected to be useful in understanding deposit growth in a variety of other systems as well.

597213 Theoretical Studies on Poisoning of Ni Catalyst in Methane Steam Reforming

Abstract:

Poisoning of catalysts due to carbon-based species is a major impediment to the productivity of industrial operations. Chemical industries spend billions of dollars annually to replace or regenerate deactivated catalysts. Ni catalysts used in Methane Steam Reforming (MSR) are highly susceptible to carbon poisoning due to the formation of graphite layers. Although numerous experimental investigations have been carried out, there is a lack of comprehensive understanding on the carbon poisoning of Ni at the molecular level. It is thus imperative to gain deeper insight into poisoning mechanisms to reduce the rate of deactivation of Ni catalysts.

Graphene (a single layer of graphite) is a representative model for carbon-based catalyst poisons. Previous studies have shown that graphene binds on Ni(111) in a commensurate manner under the influence of van der Waals forces. However, the detailed mechanism of graphene formation (within the MSR pathways) is unclear. We aim to develop a computational model that can adequately capture the MSR reaction kinetics including graphene formation and the van der Waals interaction between Ni and graphene.

In this work, spin-polarized Density Functional Theory (DFT) calculations have been employed to understand the formation of graphene in the MSR reaction. The binding energies of graphene and species of MSR – CO, C, H2O, H, O and OH – were computed using different DFT functionals. The latter include PBE and RPBE (GGA functionals), PBE-D3 and RPBE-D3 (GGA functionals with a posteriori dispersion corrections), optB86b-vdW and optB88-vdW (van der Waals DF). The predictions of these functionals have been benchmarked to experimental data in a systematic way. The PBE functional exhibits good agreement with experimental data for most species of MSR; however, it does not capture the van der Waals interactions between graphene and Ni. RPBE and RPBE-D3 predict the binding energy of CO reasonably well compared to other functionals. Overall, PBE-D3 has been found to predict binding energies of MSR species and graphene with acceptable accuracy.

Subsequently, transition state calculations have been performed (using the PBE-D3 functional) to estimate the activation barriers of different MSR reaction pathways. The three main MSR reforming pathways have comparable activation energies at the electronic level. A KMC model can help us discern the dominant reforming pathway at MSR conditions. These computational investigations can provide qualitative insights to control and mitigate the poisoning rate of Ni catalyst at MSR conditions.

610653 Surface Acidity Characterization of Metal-Doped Amorphous Silicates Via Py-FTIR and ^{15}n NMR Simulations

Abstract:

Metal-doped amorphous silicates are effective acidic catalysts for numerous industrial applications, including dehydration of alcohols, hydrodeoxygenation of biomass, and metathesis of olefins. The acidity of these catalysts is derived from the metal dopant and is experimentally characterized using a number of techniques, such as ammonia temperature-programmed desorption (NH_3-TPD), pyridine Fourier transform infrared (Py-FTIR), and ^{15}N nuclear magnetic resonance (NMR) spectroscopy of either probe molecule. Although these measurements provide information regarding the amount and strength of the acid sites, they alone do not provide any detail about the local structure around the metal center, which hampers the ability to systematically design better catalysts. In this study, we present density functional theory (DFT) simulations of Py-FTIR and ^{15}N NMR spectra to characterize the local structure and acidity strength of metal-doped silicate clusters. We perform calculations on the model structures compiled in the METal-doped Amorphous SIlicate Library (METASIL), which aims at providing realistic amorphous structures for Zr-, Nb-, and W-doped silicates. By matching the calculated normal mode frequencies with experimental IR spectra, we reproduce the differentiation between Lewis and strong Brønsted acid sites. However, our simulations show that the coordination of Py to weak Brønsted acids is indistinguishable from that to metal sites, as they elicit similar frequency shifts. On the other hand, ^{15}N NMR can distinguish between Py interacting with metal centers, weak and strong Brønsted acid sites, and it may be able to identify the structure of metal dopants. Therefore, these results can be used to provide theoretical support for the design of new catalytic materials.

605872 A DFT Study for the Mechanistic Understanding of Passive NO_x Adsorption on Pd/H-BEA Under Dry Condition

Abstract:

With constantly changing fuel sources and improvements in engine technology the treatment of vehicle emissions at low temperature poses a formidable challenge. Although there has been a success in addressing NO_x emissions at temperatures above 200°C with the aid of selective catalytic reduction (SCR) technology, emission control during vehicle cold start (temperatures < 200°C) remains arduous. Metal-exchanged zeolites functioning as passive NO_x adsorber have been proposed as a possible solution. Their role is to adsorb NO_x from the engine exhaust at low temperature and release them when the downstream SCR reaches operating temperature.

In the present work, we have used density functional theory to gain a mechanistic understanding of NO_x adsorption on Pd/H-BEA, and have proposed a sequence of feasible elementary steps that can explain observations from flow reactor experiments for NO_x trapping.[1] Due to the complex nature of palladium speciation in zeolite pores, the structure of the active site remains elusive. For zeolites with atomically dispersed Pd in the absence of water the most likely active sites are [Z_2Pd^{II}], [$ZPd^{II}OH$], and [ZPd^{I}]. On the [Z_2Pd^{II}] site NO can be stored by reversible molecular adsorption or in the form of nitrate (Figure 1). The [$ZPd^{II}OH$] site catalyzes low energy pathways for NO and also CO oxidation, while [ZPd^{I}] can be dynamically formed and provides the strongest NO binding site. The presence of CO as stronger reductant was found to be beneficial for NO_x storage. The active site assignments are corroborated by vibrational analysis and comparison to observed experimental diffuse reflectance spectra.[1]

Capturing the temperature dependent evolution and role of active sites in Pd/H-BEA upon NO exposure will expedite the development of suitable PNA materials to achieve desired emission control performance at low temperature.

Reference:

1. Malamis, Sotirios A., Michael P. Harold, William S. Epling, *Industrial & Engineering Chemistry Research* 58,22912-22923 (2019).

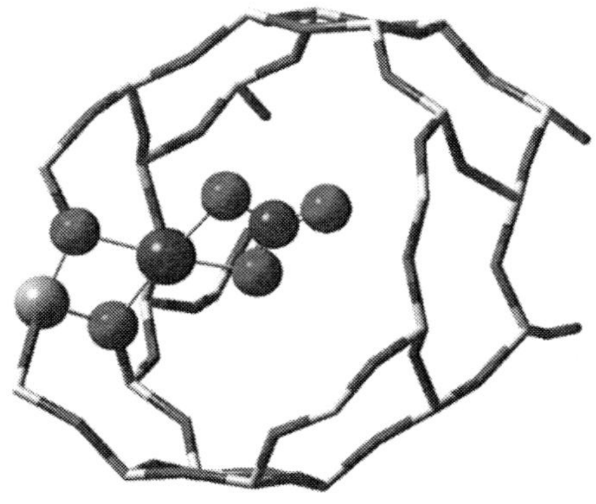

Figure 1: Bidentate nitrate formation on a naked [Z_2Pd^{II}] active site in BEA zeolite. Red, blue, sand, teal, and reddish gray atoms represent O, N, Si, Pd, and Al, respectively.

604132 Deposition and Structure of MoO_3 Clusters on Anatase $TiO_2(101)$

Abstract:

Oxide clusters supported on metal oxide substrates are important as heterogeneous catalysts[1, 2] due to their potential tunability: producing variable cluster-support interactions with a range of chemically- and photochemically-active structures.[3] Here, the behavior of MoO_3 clusters deposited on a well-defined anatase TiO2(101) surface has been explored via scanning tunneling microscopy (STM), X-ray Photoelectron Spectroscopy (XPS), and density functional theory (DFT) calculations including ab initio molecular dynamics (AIMD) simulations. From this work, we are able to deduce the constraints on the MoO_3 cluster's structure, distribution, and redox properties as a function of cluster coverage and temperature, which, now known, will indicate the reactive properties one can expect from this material.

Monodispersed MoO_3 clusters are deposited by the sublimation of MoO_3 powder at 1350 K over the anatase $TiO_2(101)$ surface. Since the stable gas phase structure of MoO_3 is a trimer in cyclic configuration, i.e. $(MoO_3)_3$,[1, 4] it is assumed that MoO_3 deposits as this structure. After deposition, the STM images of the lowest concentration of MoO_3 show no preference for step edges, but three-unit clustering is observed. The adsorbed clusters appear as bright protrusions (see Figure 1a), with an apparent cluster height of approximately 1.5 Å. Annealing to 450 K results in a better-ordering of the overlayer, but further annealing to 650 K leads to three-dimensional clusters. The XPS results indicate that the $Mo(3d_{5/2})$ binding energy in as-deposited $(MoO_3)_3$ is characteristic of Mo^{6+}, and the oxidation state of Mo remains (+6) upon heating to 500 K, consistent with DFT results suggesting that the O of MoO_3 monomers are very tightly bound.

AIMD simulation of the most stable adsorption configuration of terrace-bound $(MoO_3)_3$ reveals that this structure quickly decomposes into monomeric units of MoO_3. This decomposition is shown to be primarily entropically driven. As can be seen in Figure 1(a-b), STM imaging clearly shows that these monomeric units preferentially remain in one of four as-decomposed trimeric orientations—the three-unit clusters seen in STM—elucidated via DFT to be isoenergetic structures labeled A-D in Figure 1. DFT calculations show isolated monomers to be energetically (and naturally, entropically) preferred over these trimeric orientations. However, activation barriers for monomer MoO_3 diffusion are calculated to be significant. The along-row diffusion barrier is shown to be thermally surmountable while the across-row diffusion barrier is prohibitively high. This effectively constrains MoO_3 monomers to remain in the A-D trimeric orientations provided high temperatures are not accessed, at which point they are constrained to a single anatase $TiO_2(101)$ row, similar to that found for adsorbed H.[5] As such, this system may offer great promise as an ideal platform for reactivity studies on well-defined supported model transition-metal oxide catalysts.

References:

1. Rousseau, R., et al., *Chemical Society Reviews*, 2014. **43**(22): p. 7664-7680.
2. Bondarchuk, O., et al., *Angewandte Chemie International Edition*, 2006. **45**(29): p. 4786-4789.
3. Tang, X., et al., *Journal of Physics: Conference Series*, 2013. **438**: p. 012005.

4. Berkowitz, J., M.G. Inghram, and W.A. Chupka, *The Journal of Chemical Physics*, 1957. **26**(4): p. 842-846.
5. Doudin, N., et al., *ACS Catalysis*, 2019. **9**(9): p. 7876-7887.

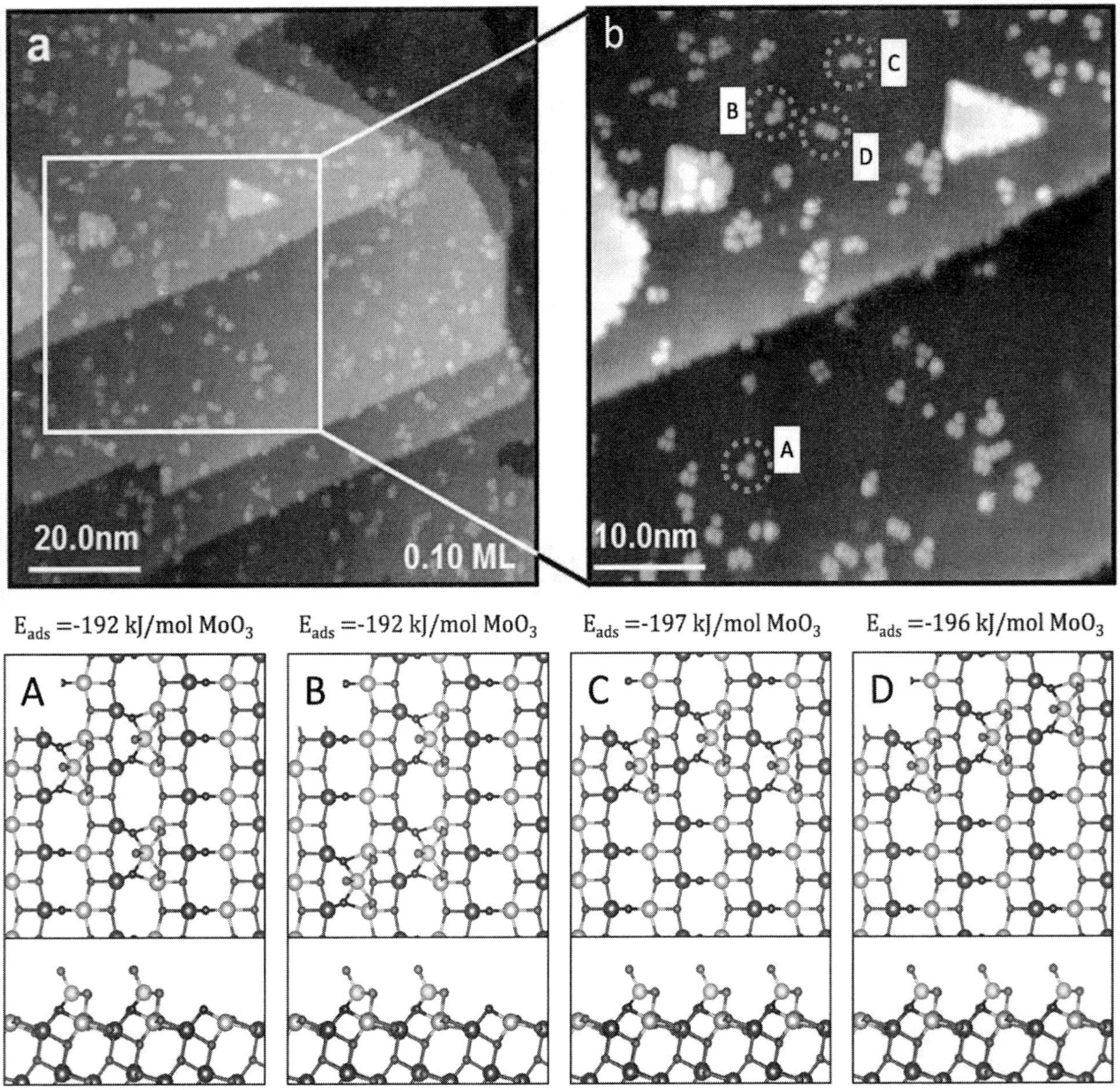

Figure 1. (a) STM imaging of MoO_3 on anatase $TiO_2(101)$, deposited to 0.10 ML MoO_3 coverage via sublimation of MoO_3 powder, with a (b) zoomed image of the surface showing the preponderance of trimeric orientations of the as-deposited MoO_3. The individual MoO_3 adsorbates are shown to be monomeric. Four trimeric orientations dominate, labeled A-D, elucidated via DFT to be in the adsorption conformations shown (turquoise spheres are Mo; magenta spheres are the O of MoO_3; purple and red spheres are 2-fold and 3-fold coordinated surface O of anatase $TiO_2(101)$, respectively; while dark and light grey spheres are 6-fold and 5-fold coordinated Ti of anatase $TiO_2(101)$, respectively.

607916 Interconversion-Controlled Liquid-Liquid Phase Separation of a Molecular Chiral Model

Abstract:

Liquid-liquid phase separation of fluids exhibiting interconversion between alternative states is a ubiquitous phenomenon, which can explain fluid polyamorphism and may impact biological function. However, molecular-level insight on the interplay between competing forces that drive and resist phase separation in interconverting fluids remains elusive. Here, we utilize a three-dimensional off-lattice model of enantiomers with tunable chiral interconversion and interaction properties, to elucidate the essential physics underlying the stabilization and tunability of phase separation in fluids with interconverting states. We show that quenching such a system below the liquid-liquid critical temperature results in demixing via spinodal decomposition and, simultaneously, mixing via chiral interconversion. The competition between negative-diffusion demixing and interconversion-driven remixing results in micro-phase phase separation, kinetically arrested by the interconversion of enantiomers. We have also found that the competition and mutual coupling between the enantiomer interconversion and diffusion make the effective diffusion coefficient to be strongly dependent on the interconversion rate and the size of micro-phase domains. These findings provide novel insights on how the interplay between kinetics and thermodynamics define the steady-state morphology of phase transitions in fluids with interconverting molecular or supramolecular states.

609324 Solubility Measurement and SAFT-γ Mie Prediction of Peptides Based on Amino Acid Residues

Abstract:

Peptides are short chain biomolecules containing 50 or fewer amino acids, which are known to have important therapeutic properties and huge potential application in the treatment of chronic and metabolic diseases (e.g. cancer, obesity and diabetes).[1-4]Peptide crystallisation, as a good alternative to chromatographic purification, can solve the shortcomings of traditional purification method, such as high cost, proteolytic degradation and physiochemical instability. Accurate solubility data is a crucial thermodynamic property during the crystallisation, as corresponds to the stable phase boundary in the phase diagram, it is needed to characterize supersaturation, calculate nucleation parameters in the prediction and design of crystallisation conditions. Currently only limited solubility data of peptides in organic solvent systems is available in the literature.

We present a study of the chain length and side chain effect on the thermodynamic properties of several peptides and homopeptides during crystallisation to establish a rational design of the conditions for glycine homopeptide crystallisation. The solubility of glycine homopeptides (glycine, diglycine, triglycine, tetraglycine, pentaglycine and hexaglycine), amino acids with different side chains (aspartic acid, phenylalanine, histidine and tyrosine) and their dipeptides (asp-phe, gly-asp, gly-phe, phe-phe, gly-gly, tyr-phe, gly-tyr, gly-his) in water from 278.15K to 313.15K were measured using the gravimetric method.

Compared with small molecules, the solubility measurement of longer chain peptides is challenging because of their flexible structure, especially in the case of the peptides composed of more than five amino acid residues. During the crystallisation process, the adding of the salts and precipitants which stabilise the nature and the structure of the peptides also makes the measurement of solubility become more complex. In this context the development of tools to deliver accurate predictions of solubility of peptides from limited experimental data is also of critical interest.

The SAFT- γ Mie group contribution equation of state in which a heteronuclear model is implemented, with molecules split into their chemical moieties (CH_3, CH_2, NH_2, COOH, etc...) to calculate the thermodynamic properties and phase equilibrium (including solid-liquid solubility)[5] of fluids predictively. We model the peptides and amino acids of interest divided into the relevant groups. Known group-group interaction parameters are adopted from the literature or developed in this work using literature data as well as data measured in this study. The NH_2 -COOH parameters are determined based on the thermodynamic properties of glycine, alanine and valine. These parameters can be used to predict the properties of other amino acids (alanine, valine, aspartic acid, glutamic acid, isoleucine, leucine and lysine) and peptides.

In summary, the effect of chain length and side chain on the solubility of peptides is studied and the group-group interaction matrix of the SAFT-γ Mie approach is extended for the prediction of the solubility of amino acids and peptides. This research establishes a thermodynamic foundation for the further design of peptides crystallisation process and explores the application of SAFT-γ Mie to biomolecular thermodynamic properties, providing a flatform to explore the relationship between their physicochemical properties and biostructures.

1. Chaparro-Aguirre, E.; Segura-Ramirez, P. J.; Alves, F. L.; Riske, K. A.; Miranda, A.; Silva Junior, P. I., Antimicrobial activity and mechanism of action of a novel peptide present in the ecdysis process of centipede Scolopendra subspinipes subspinipes. *Sci Rep* **2019,** *9* (1), 13631.
2. Jeong, W. J.; Bu, J.; Han, Y.; Drelich, A. J.; Nair, A.; Kral, P.; Hong, S., Nanoparticle Conjugation Stabilizes and Multimerizes beta-Hairpin Peptides To Effectively Target PD-1/PD-L1 beta-Sheet-Rich Interfaces. *J Am Chem Soc* **2020,** *142* (4), 1832-1837.
3. Pittala, S.; Levy, I.; De, S.; Kumar Pandey, S.; Melnikov, N.; Hyman, T.; Shoshan-Barmatz, V., The VDAC1-based R-Tf-D-LP4 Peptide as a Potential Treatment for Diabetes Mellitus. *Cells* **2020,** *9* (2).
4. Tomaro-Duchesneau, C.; LeValley, S. L.; Roeth, D.; Sun, L.; Horrigan, F. T.; Kalkum, M.; Hyser, J. M.; Britton, R. A., Discovery of a bacterial peptide as a modulator of GLP-1 and metabolic disease. *Sci Rep* **2020,** *10* (1), 4922.
5. Papaioannou, V.; Lafitte, T.; Avendano, C.; Adjiman, C. S.; Jackson, G.; Muller, E. A.; Galindo, A., Group contribution methodology based on the statistical associating fluid theory for heteronuclear molecules formed from Mie segments. *J Chem Phys* **2014,** *140* (5), 054107.

610301 Bridging the Gap between Static Lattice Energy and Full Free Energy Differences of Organic Crystal Polymorphs When Predicting Crystal Energy Landscape

Abstract:

To reduce the risk of unknown polymorphs showing up in late stage development, crystal structure predictions (CSPs) have been employed to determine possible forms. Due to the fact that crystalline structures of small organic molecules can vary in properties, fields such as pharmaceutical formulation, organic electronics processing and explosives preservation have interest in fast and efficient methods to predict favorable structures. A failure to know all potential solid forms has led to recalls, reformulation, or a missed opportunity to utilize a material to its fullest extent. CSPs can be used to determine the solid energy landscape *a priori* by exhaustively generating solid forms and ranking them based on their relative energies.

Ranking of crystals in a CSP have historically relied on the lattice energy, but more recently free energy techniques have been introduced to help capture crystal reranking due to temperature changes. Thermodynamic ranking of the crystal structures allows one to determine the likely crystal structures to find experimentally. A CSP can generate 50-200 crystal structures within 2.5 kcal/mol of the global minimum so accurate ranking is critical when it's likely that only a handful will actually crystallize. Lattice energy calculations have been the most widely used technique to rank crystal structures, but polymorph stability is known widely to be temperature dependent and therefore entropic effects cannot be neglected. Phonon calculations have been integrated to determine harmonic free energies for CSPs and, while not published, we are aware of free energy calculations using full molecular dynamics (MD) in industrial settings.

The required level of accuracy in free energy calculations for CSPs is unknown, which is why we have developed a whole host of methods in varying accuracy to compute the solid-solid free energy differences. MD is the gold standard, allowing us to compute a full thermodynamic cycle between two crystals for a given potential, but is fairly expensive to conduct on upwards of 200 crystals. On the opposite side of the spectrum are lattice dynamics techniques that use the static lattice phonons of the crystal to determine the entropic contributions to the free energy. Lastly, we have developed an intermediate technique to include the full dynamic behavior of molecular dynamics with the speed of compute the crystal phonons to come up with an approximation to the free energy that introduces error of 0.3 kcal/mol. I will discuss the specifics of these methods and compare their relative speed, accuracy, and direct implementation into a CSP to help screen for likely solid forms of molecules of interest.

611451 Prediction of Physical Properties and Phase Behavior for Future Jet Fuel Selection Via Monte Carlo Simulations

Abstract:

Our fuel sources will change over time as we transition to different raw fuel sources, strive to reduce emissions, and develop renewable fuel blends. These changes will drive the automobile and aviation industries to evaluate their fuel sources and fuel blends. The industries will adapt to these changes and environmental concerns, and the aviation industry shall develop new fuel alternatives and fuel blends. If these new fuel combinations can reduce the overall energy consumption or the emissions even by a few percentage points, the overall sum of these savings will be quite large.

Liquid fuels used in internal combustion and jet engines contain hundreds of species, and therefore computational modeling efforts of combustion dynamics have focused on fuel "surrogates" that mimic the physical and combustion properties of the fuel of interest. These surrogates are composed primarily of linear and branched alkanes containing 8 to 17 carbon atoms, as well as cyclic molecules, such as toluene and decalin.

This work validates the use of the current molecular models (force fields), such as the Transferable Potentials for Phase Equilibria (TraPPE) for reproducing physical properties (liquid density, surface tension, and heat of vaporization), and vapor-liquid phase diagrams of typical jet fuel surrogates (e.g., JP-8, Jet-A). Properties are predicted from Monte Carlo simulations using the GOMC[1] simulation package, with workflows created using the Molecular Simulation Design Framework (MoSDeF)[2-5], which automatically generates this output with minimal command input. The combination of accurate molecular models, high-performance simulation engines, and automated workflows are expected to enable the use of Monte Carlo simulations for testing future jet fuel combinations while minimizing the overall cost of research by narrowing down the required experimental testing to a few candidates.

[1] Nejahi, Y, Barhaghi, MS, Mick, J, Jackman, B, Rushaidat, K, Li, YZ, et al. *GOMC: GPU Optimized Monte Carlo for the simulation of phase equilibria and physical properties of complex fluids.* Softwarex, 2019; **9**: 20-7.

[2] Klein, C.; Sallai, J.; Jones, T. J.; Iacovella, C. R.; McCabe, C.; Cummings, P. T. "In Foundations of Molecular Modeling and Simulation: Select Papers from FOMMS 2015"; Springer: Singapore, 2016.

[3] Klein, C.; Summers, A. Z.; Thompson, M. W.; Gilmer, J. B.; McCabe, C.; Cummings, P. T.; Sallai, J.; Iacovella, C. R. "Formalizing atom-typing and the dissemination of force fields with foyer". Computational Materials Science 2019, 167, 215-227.

[4] Cummings, P. T.; Gilmer, J. B. Open-Source Molecular Modeling Software in Chemical Engineering. Current Opinion in Chemical Engineering 2019, 23, 99-105.

[5] Jankowski, E.; Ellyson, N.; Fothergill, J. W.; Henry, M. M.; Leibowitz, M. H.; Miller, E. D.; Alberts, M. e. a. Perspective on Coarse-Graining, Cognitive Load, and Materials Simulation. Computational Materials Science 109129, 171, 2020.

600441 Stabilizing Blue Phase Liquid Crystal Under Spherical Confinement

Abstract:

At high level of chirality, cholesteric liquid crystals self-organize their helical structure into double twisted cylinders, which are interspersed by line defects. The double twisted cylinders organize into three-dimensional cubic lattice structures that can reflect the incident light in the wavelength range of visible spectrum. These structures, that emerge between cholesteric and isotropic phases, are known as cholesteric "Blue Phases, BPs". The blue phase liquid crystals are, however, thermodynamically stable over a narrow temperature range. Introducing reactive mesogens into the defect structures followed by polymerization has been proven as a successful strategy to expand the stability temperature of BPs. While the bulk properties of the polymer-stabilized blue phases have extensively been explored, the effect of geometrical confinement on the stabilized blue phases remains as an open question. Our experimental and computational studies show that confining the polymer-stabilized blue phases within the spherical droplets significantly influences their internal structures, reflected color and stability temperature. Using a super-resolution confocal laser scanning microscopy, we have been able to image the periodic structure of the double twisted cylinders in real space. We have found that the orientation of the lattice structures in the stabilized BP droplets can be manipulated by controlling the droplet size and polymerization time, which can be exploited to achieve monodomain blue phase structures. These new findings offer tremendous opportunities for engineering optical devices based on the high chirality liquid crystals.

605515 Fluid Filling Impedes Thermal Conduction in Free-Standing, Single-Walled Carbon Nanotubes

Abstract:

Carbon nanotubes are an indispensable testbed for materials properties at the quasi 1D limit, and hold significant promise for use as thermal devices and nanofluidic conduits. While they enable exceptional heat and mass transfer, the interplay between these two of their attributes is largely unexplored territory. Here we use Raman spectroscopy to elucidate the effect of fluid filling on a nanotube's thermal conductance. We study 17–200 μm long, free-standing segments of individual nanotubes in vacuum, reducing heat transfer pathways to the minimum. We show that a simple heat transfer equation based on Fourier's law can be applied and extract an average thermal impedance of ~1x10^-3 K m/W due to water filling of a (17,9) nanotube. Our findings support confined fluids to enhance phonon scattering and thereby impede thermal conduction in these and related nanochannels.

608070 Phase Transitions of Water Confined in Nanopores

Abstract:

Phenomena that occur only at very high pressures in the bulk phase are often observed to occur in the confined phase at normal pressures. Abundant experimental evidence suggests that adsorbates confined in nanoporous carbons exhibit high pressures, such as high-pressure crystal structures, high-pressure chemical reactions, and the deformation of pore walls due to the adsorbate. Also molecular simulation studies of the pressure tensor for simple adsorbates in carbon nanopores of slit, cylindrical and spherical geometries show that for modest bulk phase pressures the pressure parallel to the pore walls is of the order of GPa, while the pressure normal to the wall is of the order of hundreds of MPa [1,2].

We report X-ray, neutrons diffraction (ND) and dielectric studies of ice confined in activated carbon fibres (ACFs). The fibres are built of turbostratic nanoparticles separated by quasi two-dimensional voids, forming narrow slit-shaped pores. The structure of ice in cylindrical nanocarbons pores (CMK-3), which are the reverse carbon replica of silica SBA-15 matrices, was also analyzed. The results show strong evidence for the presence of both hexagonal and cubic forms of ice confined in all of the systems studied. We were not able to obtain a good Rietveld refinement of our diffraction patterns with typical hexagonal and cubic forms of ice; but we observe that the formation of the ice structure in the pores, has been attributed to the restrictions on the size of the crystallite and the quasi-high pressure effect in pores. However, according to recently published work [3], the cubic form of ice is not exactly cubic, but contains stacking faults instead of an ordered arrangement of two dimensional layers. This form of ice is in fact a stacking-disordered material containing cubic sequences interlaced with hexagonal sequences. The structure of the confined ice observed in our experiments in nanocarbons pores corresponds to disordered stacking ice Isd,, which was identified as having the space group P3m1 [3,4]. We observe,that the stacking disorder can vary in complexity depending on the nature of the pore walls and on the prevailing thermal conditions during the ice formation. An analysis of the kind of I_{sd} formed in ACF and CMK-3 of different pore sizes will be presented. These crystal forms, which occur in bulk water only at temperatures below 180 K in the case of cubic ice, and at pressures of hundreds of MPa are stabilized by the confinement.

Financial support for the NCN grant No UMO-2016/ 22/A/ST4/00092 is gratefully acknowledged

[1] Y. Long, J.Palmer, B.Coasne, M.Sliwinska-Bartkowiak, K.E.Gubbins, Phys. Chem. Chem. Phys , 13 , 17163, (2011)

[2] M.Sliwinska-Bartkowiak, M. Drozdowski, M. Kempiński, Y. Long, J. Palmer, K.E. Gubbins, Phys. Chem.Chem.Phys.,14,7145 (2012);

[3 J.C.Palmer, F.Martelli, Y. Liu, R. Car., A.Z. Panagiotopoulos, P.G. .Debenedetti, , Nature, 510 , 385 (2014); T. Malkin, B. J. Murray, A. Brukhno, J. Anwar and C. G. Salzmann ,Proc. Natl. Acad. Sci. USA, 109 ,1041(2012)

[4] K.Domin, K-Y. Chan, H. Yung, K.E. Gubbins, M. Sliwinska-Bartkowiak,,J.Ch.E.D., 61 (2016), J.Mol. Liq., 43,1580 (2020)

608060 Vapor-Liquid Equilibrium Data for R32, R134a, and R1234ze(E) Systems over the Temperature Range from (263 to 323) K

Abstract:

Hydrofluoroolefins (HFOs) are of prevailing interest as they may serve as environmental-friendly alternatives to replace chlorofluorocarbons (CFCs) and hydrofluorocarbons (HFCs) in numerous applications. For the design and simulation of refrigeration systems using mixtures containing HFOs, precise knowledge of their thermophysical properties, especially vapor-liquid equilibrium (VLE) data, is essential. However, based on a literature survey, the public experimental data for R32, R134a, and R1234ze(E) systems are still limited.

For this research, a liquid-recirculation analytical apparatus was developed to carry out the measurements (FIG. 1).[1] The VLE experimental data for the binary mixtures of R32 + R134a, R32 + R1234ze(E), R134a + R1234ze(E), and the ternary system of R32 + R134a + R1234ze(E) were obtained over the temperature range from (263 to 323) K. The standard uncertainties of the temperature and pressure measurements are 10 mK and 0.5 kPa, respectively. The standard uncertainties of both vapor and liquid phase mole fractions are estimated to be within 0.005. The binary VLE properties were correlated by the Peng-Robinson-Stryjek-Vera equation of state[2] combined with the Wong-Sandler (WS) mixing rule and the non-random two-liquid activity coefficient model (NRTL). With the fitted parameters of binary systems from this work and the literature, the ternary VLE properties were predicted. Good agreement between the experimental data and the calculated results were obtained.

References

[1] Hu X.Z., Meng X.Y., Wu J.T., *Fluid Phase Equilib.* 431 (**2017**) 58-65.

[2] Stryjek, R., Vera, J. H., *Can. J. Chem. Eng. 64* (**1986**) 820-826

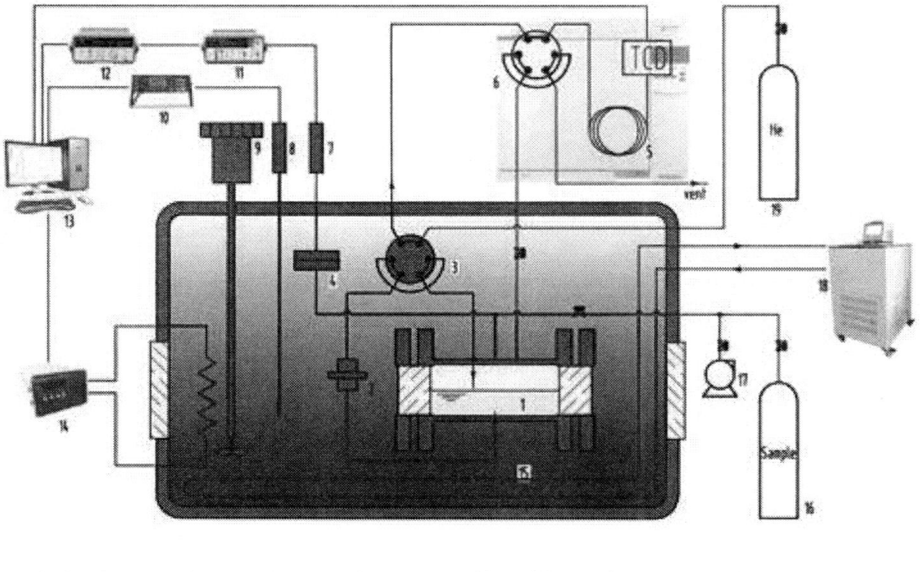

FIG. 1. Schematic diagram of the liquid-recirculation analytical VLE system.

600022 Computational Study of Ground State Organic Chemistry Reactions in the Presence of Toxic and Non-Toxic Solvents

Abstract:

Replacing toxic solvents used to carry out organic chemistry reactions with less toxic "green" alternatives would have major implications for the sustainability of chemical production. Indeed, many of the solvents currently used in the chemical industry are hazardous to human health and detrimental to the environment [1]. In this work, we investigated the effect of different types of toxic and non-toxic solvents on the kinetics of nucleophilic substitution and addition organic chemistry reactions. Reaction rate constants were computed with transition state theory by using density functional theory (DFT), string methods, and eigenvector following. In order to evaluate the possible participation of the solvent in the reaction, the presence of solvents was described by the implicit COSMO model with and without the addition of an explicit solvent molecule. Our results confirm the importance of including the solvent and provide an initial data set to guide the prediction of reaction rate constants in solvents.

[1] P. R. Koteswararao, S. L. Tulasi, and Y. Pavani, "Impact of Solvents on Environmental Pollution," *J. Chem. Pharm. Sci.*, 3:32–135 (2014).

599322 Transition-Metal Solvated-Electron Precursors- a Study of the Electronic Structure and Periodic Trends of Early Transition-Metal Ammonia Complexes

Abstract:

The addition of metals in ammonia solution form metal ammonia complexes surrounded by solvated electrons. The saturation of these solutions with metals generates a non-typical form of metalloid materials called liquid or expanded metals. These consist of a grid of positively charged metal ammonia complexes and free electrons moving around them. In our group we investigate the electronic structure of these metal ammonia units. These complexes form solvated electron precursors (SEPs) involving displacement of inner metal electrons to diffuse, peripheral super-atomic orbitals. Low-lying excited states of these complexes involve population of the 1s, 1p, and 1d outer orbitals. Currently, the ground and excited states of transition metal-ammonia complexes of $X(NH_3)_6^{0,1+}$ (X = Zr, Nb, Cr, Mo) are investigated with ab initio electronic structure theory. This work builds upon previous studies to identify the effect of the transition metal identity- illustrating a periodic trend in the electronic and chemical structures of these complexes. Further work investigates the use of polydentate amine ligands such as diethyl amine to stabilize the formation of SEPs.

606349 Exploring Metal-Support Interactions in Catalysis with Statistical Learning

Abstract:

We apply statistical learning (SL) techniques to identify descriptors that can construct models for predicting metal-support interactions in catalysis that results from charge transfer. Charge transfer between the metal and the support can be tuned by substituting parent metals in the oxide surface with aliovalent dopants or by introducing co-adsorbates on the oxide surface. As a test case, we use Al, B, Li, and Na as dopants and F, H, OH, and NO_2 as adsorbates to demonstrate control of charge transfer between transition metal atoms and modified MgO(100) surfaces. We collect fundamental chemical properties of the constituent elements in the system to generate a tremendous pool ($\sim 10^6$) of possible physical descriptors via feature engineering, which are then used to construct predictive models for computing metal binding energies on the surface. SL methods, e.g., LASSO, Horseshoe prior, and Dirichlet-Laplace prior, are then trained against DFT data to select descriptors that predict metal adatom binding energy on the MgO surface. We find that the Bayesian feature selection (FS) methods, i.e., Horseshoe prior and Dirichlet-Laplace prior, yield models that are generally more accurate than those identified with the LASSO method. Furthermore, we show that Dirichlet-Laplace prior can identify features that are transferable to other relevant oxides that are not included in the training process, such as CaO(100), BaO(100), and ZnO(100). This work demonstrates the superior capability of Bayesian FS methods for constructing general models for describing the metal-support interaction.

607476 Optical and Electronic Properties of Defective Copper Selenide Predicted from Density Functional Theory

Abstract:

Plasmonic nanomaterials have been shown to have exciting applications in catalysis and medicine. Most plasmonic materials are precious metals such as Au and Ag. There is interest in developing non-precious metal plasmonic substrates due to lower material costs. Copper chalcogenide nanoparticles are very promising materials because their optical properties can be tuned. We investigate the optical and electronic properties of copper selenide, $Cu_{2-x}Se$, through density functional theory (DFT). $Cu_{2-x}Se$ is only plasmonically active in the oxidized state, i.e., for $x > 0$. Moreover, it is known to be superionic due to the mobility of the Cu atoms. We have used both generalized gradient approximation (GGA) and meta-GGA functionals to compute structural, electronic, and optical properties of cubic $Cu_{2-x}Se$, for $x = 0$, 0.25, 0.5, and 0.75. We performed ab-initio molecular dynamics simulations to simulate disorder within the lattice, taking 50 snapshots sampled from each simulation and relaxing them to their local ground states. We used the HSE06+U meta-GGA functional to calculate electronic structures, including optical band gaps. We found that the average optical gap increases with increasing oxidation, while electronic band gaps were only observed for the $x = 0$ structure. We observed that ionic disorder induces unpredictable changes in the optical gaps of $Cu_{2-x}Se$ for each fixed value of x.

607636 Influence of Oxygen Coverage on the Dominant Routes Towards Furanic Oxidation

Abstract:

Electrochemical conversion of biomass derived furanics to industry relevant chemicals show strong promise in the area of renewable chemistry. But the understanding of the mechanism on Pt catalysts is lacking. Using a simple approach to calculation of reaction energy and potential dependent electrochemical activation barriers, it is now possible to compute the free energy landscape of complex furanic conversion pathways. We have earlier showed with Density Functional Theory (DFT) calculations that decarboxylation of furoic acid (C-C cleavage) limits the rate of furfural oxidation which seems to consistent with high selectivity towards furoic acid at moderate potentials (0.8V-1V-RHE)[1]. However modeling the reaction energetics at higher potentials is complicated with increasing coverage of oxygen on catalytic surface. Experiments show increasing selectivity towards oxidation products deeper than furoic acid at higher potentials on Pt (111). This is hypothesized to occur due to possibility of additional pathways to furanic ring oxidation facilitated through the increasing coverage of oxygen at higher potentials on the modelled clean Pt (111) surface. Energetics of decarboxylation at partial oxide layer of PtO (100) and other thermodynamically feasible oxide models do not explain the increased rate of decarboxylation at higher potentials. To investigate the effect of partial oxygen coverage for the reaction of furanics on Pt (111), temperature-programmed desorption (TPD), high-resolution electron energy loss spectroscopy (HREELS), and density functional theory (DFT) calculations were conducted. HREEL spectra and DFT calculations indicated that furan initially reacted through oxygen addition to the ring to produce surface-adsorbed furanone intermediates. HREELS and DFT calculations suggested that alcohol oxidation proceeded through dehydrogenation at the hydroxyl and methylene groups and subsequent oxidation to produce a carboxylate intermediate [2]. These results suggest that direct insertion of oxygen into the furanic ring is an energetically accessible pathway at different stages of oxidation. Insertion of oxygen/hydroxyl into furanic ring prior to C-C cleavage can allow access to reaction pathways explaining synthesis of hydroxyfuroic acid in furfural electrochemical oxidation on Pt [3] and might allow us to explain the shift in increased rate of furoic acid oxidation at higher potentials through oxygen insertion.

References:

1. Gong, L., Agrawal, N., Roman, A., Holewinski, A., & Janik, M. J. (2019). Density functional theory study of furfural electrochemical oxidation on the Pt (1 1 1) surface. Journal of catalysis, 373, 322-335.
2. Mark, L. O., Agrawal, N., Román, A. M., Holewinski, A., Janik, M. J., & Medlin, J. W. (2019). Insight into the oxidation mechanism of furanic compounds on Pt (111). *ACS Catalysis*, *9*(12), 11360-11370.
3. Roman, A. M., Hasse, J. C., Medlin, J. W., & Holewinski, A. (2019). Elucidating Acidic Electro-Oxidation Pathways of Furfural on Platinum. *ACS Catalysis*, *9*(11), 10305-10316.

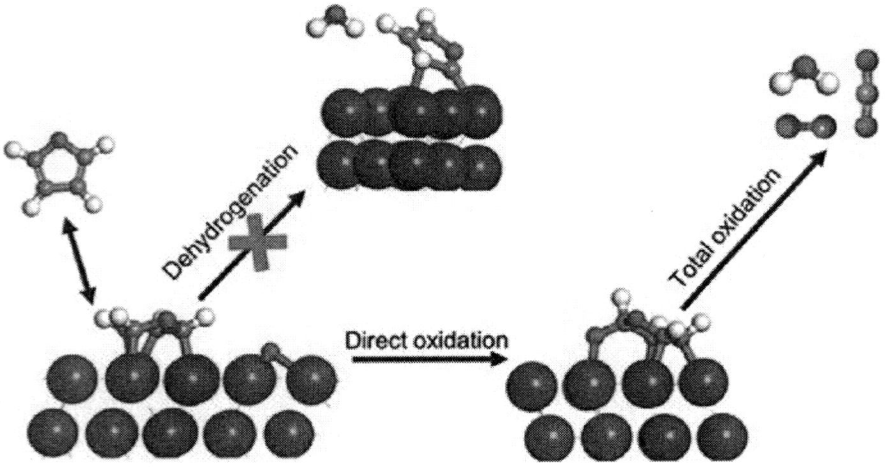

607733 Oxide-Generation Reactions in Lithium-Ion Batteries: A Reaction Coordinate Analysis

Abstract:

In lithium-ion batteries, oligomeric species have been experimentally detected in the solid-electrolyte interphase (SEI) and are expected to form during late-stage SEI growth.[1,2] Often, the mechanisms proposed in literature posit that an oxide functional group acts as the propagator of the oligomeric growth[3]; however, the formation of the first oxide groups that initiate these chain reactions has been a topic of little focus. Here, we investigate various reactions (largely decarboxylation reactions) that lead to these reactive oxide groups and their associated reaction coordinates using the aimless shooting and likelihood maximization procedure.[4] The reaction coordinates of each elementary step grant insight into the necessary chemical environment for the process to occur and allows for the comparison of interfacial (electrode-electrolyte interface) and bulk phase (liquid electrolyte) reaction mechanisms. The reaction coordinates, coupled with associated energetic profiles, allow for the estimation of the most probable sources of reactive oxide functional groups.

References:

(1) Jin, Y.; Kneusels, N. J. H.; Marbella, L. E.; Castillo-Martínez, E.; Magusin, P. C. M. M.; Weatherup, R. S.; Jónsson, E.; Liu, T.; Paul, S.; Grey, C. P. Understanding Fluoroethylene Carbonate and Vinylene Carbonate Based Electrolytes for Si Anodes in Lithium Ion Batteries with NMR Spectroscopy. *J. Am. Chem. Soc.* **2018**, *140*, 9854–9867.

(2) Jin, Y.; Kneusels, N. J. H.; Magusin, P. C. M. M.; Kim, G.; Castillo-Martínez, E.; Marbella, L. E.; Kerber, R. N.; Howe, D. J.; Paul, S.; Liu, T.; et al. Identifying the Structural Basis for the Increased Stability of the Solid Electrolyte Interphase Formed on Silicon with the Additive Fluoroethylene Carbonate. *J. Am. Chem. Soc.* **2017**, *139*, 14992–15004.

(3) Burkhardt, S. E. Impact of Chemical Follow-up Reactions for Lithium Ion Electrolytes: Generation of Nucleophilic Species, Solid Electrolyte Interphase, and Gas Formation. *J. Electrochem. Soc.* **2017**, *164*, A684–A690.

(4) Peters, B.; Trout, B. L. Obtaining Reaction Coordinates by Likelihood Maximization. *J. Chem. Phys.* **2006**, *125*.

609994 Understanding the Role of Alkali Metals to Promote Pure Nickel and Bimetallic Nickel-Copper Catalysts for Dry Reforming of Methane

Abstract:

Dry reforming of methane (DRM) is one of the reactions in which numerous studies have been conducted to simultaneously convert CO_2 and CH_4 into syngas. Earlier work has shown that bimetallic Ni-Cu catalyst possesses better activity to form coke than conventional pure Ni (111) catalyst. It has been identified that certain promoters such as alkali metals can enhance the activity and minimize carbon deposition on the DRM catalysts. Hence, we examined whether alkali metal promoters (Na and K) promote catalytic ability of the two catalysts for DRM with the help of a first-principles study utilizing density functional theory (DFT). Two different loadings (0.3% and 3%) for the alkali metal promoters were considered and how they influence the stability of two DRM catalysts. In addition to the effects of varying promoter loadings, another factor to determine the energetic stability of the catalytic system was to study the adsorption site of the promoter atoms. This was achieved by studying the relative energy along the adsorption site.

The smallest relative energy for pure Ni (111) was achieved by Na and K atoms at 3% loading of about 8.26 eV and 8.74 eV, respectively, with the promoter atoms either preadsorbing on the top layer of the catalyst slab or replacing one of the present Ni atoms in the top layer. In addition, when two promoters were present in the bimetallic Ni-Cu catalyst exactly the same as the position located in the pure Ni (111), Na and K achieved the smallest relative energy for at -2.58 eV and -2.28 eV, respectively, and the stable catalyst configurations were obtained.

Next, we investigated the carbon sources of CH_4 and CO_2 in the DRM reaction, starting with the adsorption of C atoms at a specific adsorption site to evaluate whether two promoters on catalysts can minimize the deposition of coke. We found that K atoms at 0.3% loading promoted carbon adsorption on top sites of the pure Ni (111) catalysts, whereas for the bimetallic Ni-Cu catalysts, the carbon adsorption at the top and bridge sites was increased. Under the same loading conditions of the K promoter, Na atoms preferred adsorption of C atoms at the top and bridge sites of the bimetallic Ni-Cu system, but not in the pure Ni (111) system. With respect to the adsorption of the DRM intermediate CH_4, both Na and K atoms at 0.3% loading favored adsorption on top and bridge sites of both bimetallic Ni-Cu and pure Ni (111) catalysts.

We therefore expect that the DRM reaction occurring on the Na-promoted pure Ni (111) catalysts would give a better tendency to prevent coke formation.

610061 Nitrogen Electroreduction: Elementary Kinetic Analysis and Surface Functionalization for Improved Catalysis

Abstract:

In a future with plentiful renewable electricity, electro-chemical systems to carry out chemical transformations can provide an efficient and sustainable alternative to fossil-fuel based energy consumption. Several important chemical reactions, such as CO_2 and N_2 reduction, have electro-chemical counterparts to form useful products through a series of proton-coupled electron transfer (PCET) reactions. The activity and selectivity of PCET reactions can significantly be impacted by the electrode-electrolyte interfacial properties, and tuning this near-surface environment can be used as a control towards developing active and selective electrochemical conversions. Surface-bound amino acid chains can be used to alter the electrocatalytic performance by altering the active site structure and local environment.

We first use a simple Density Functional Theory (DFT) based approach to estimate the electrochemical barriers [1] to elucidate the elementary kinetics of the possible associative mechanisms for N_2 electroreduction (NRR) on two low index surfaces of Fe [2]. Calculated activation barriers suggest significantly larger overpotentials for NRR than those inferred from consideration of only elementary reaction free energies. Key step barriers on low index surfaces of late transition metals resulted in a "kinetic volcano" for nitrogen reduction over (111) surfaces of FCC metals. This analysis suggests that transition metal catalysts will be ineffective for NRR, due to high elementary step barriers and low selectivity [3].

We further use DFT to study the surface binding of short amino acid chains on Fe_2O_3 surfaces to understand how they alter the interfacial environment through surface-ligand interactions. We examine the effects of small amino acid binding on the stability of different Fe_2O_3 terminations in the electrochemical environment. We consider how presence of continuum solvation affects the binding preferences of the amino acid. Coverage effects on binding energies are also evaluated.

Significance

This work demonstrates the extreme challenge in developing active and selective N_2 electroreduction catalyst to produce ammonia. Also, use of amino acid as co-catalyst to modify catalyst surface and environment provides a unique design variable to improve catalytic activity and selectivity.

References

[1] G. Rostamikia et al., J. Power Sources 196(22), 9228–9237 (2011).

[2] S. Maheshwari et al., J. Chem. Phys. 150, 041708 (2018)

[3] G. Rostamikia et al. Cat. Sci. & Tech, 9 (2019) 174-181.

610171 Predicting an Optimal Oxide/Metal Interface Catalyst for Hydrodeoxygenation Chemistry of Biomass Derivatives

Abstract:

Selective C-O cleavage is the most difficult chemical transformation en route to fuels production from biomass derivatives such as furfuryl alcohol. Metal oxide-metal interfaces have recently been used to manipulate catalytic selectivity in such multistep reactions, and also hinder non-selective decarbonylation (DCO, C-C activation). Our DFT interfacial model of TiO_2/Pd core-shell catalyst in the form of rutile TiO_2 (110) nanowire over Pd (111) provides qualitative mechanistic determination of the role of interfacial active sites towards deoxygenation and decarbonylation[1]. A much-lowered barrier to hydrodeoxygenation (HDO) is obtained over TiO_{2-x}/Pd oxygen deficient interface model relative to supported Pd catalysts (0.20 eV barrier vs 0.95 eV on Pd (111)).

With this mechanistic insight, we have extended this model to predict an optimal combination of oxides and metal catalyst with interfacial properties that combine hydrogenation and redox requirements of HDO. Descriptors that dictate the synergy between the oxide and metals' functionalities for HDO at the oxide/metal interface are evaluated, including the interfacial reducibility, metal–oxygen bond strength, metal-carbon binding energy, metal's work function or its relative d-band center. We identify a greater stabilization of the C-O activation transition state through electronic charge redistribution at the interface, facilitated by a higher metal work function. Stronger metal-carbon binding dictates the favorable hydrogenation of the resulting organic fragment. The mechanistic study is also successfully extended to predict deoxygenation of other stringent oxygenates like m-cresol, phenols etc., to establish more generic correlations aimed to optimize the overall HDO chemistry, hydrogen activation and subsequent hydrogenation of different oxygenated biomass derivatives.

[1] S. Deo, W. Medlin, E. Nikolla, M.J. Janik, Reaction paths for hydrodeoxygenation of furfuryl alcohol at TiO2/Pd interfaces, Journal of Catalysis, 377 (2019) 28-40.

610685 DFT Insights into Propane Dehydrogenation Mechanisms over Iron Carbide

Abstract:

Catalytic propane dehydrogenation (PDH) has emerged as an important strategy for meeting the increasing demand for propylene caused by the shale gas revolution. The activity of Fe_3C toward PDH has been demonstrated in the literature, yet the underlying mechanism is largely unknown (e.g., Tan et al., *ACS Catal.* 6, 5673, 2016). In this work, we used density functional theory (DFT) to better understand why the PDH reaction is highly selective on Fe_3C surfaces. The stability of various Fe_3C surface terminations were identified as a function of the reaction environment using *ab initio* thermodynamics. These phases then served as our surface models for investigating kinetic barriers during the PDH reaction, where the climbing image nudged elastic band (cNEB) method was used to obtain the kinetic barriers of each reaction step. By comparing the propylene desorption barrier with the barrier for further dehydrogenation, we find that carbon-rich surfaces show much higher selectivity compared to iron-rich surfaces for propylene production over competing cracking reactions. The propylene desorption barrier was found to be critical to the overall selectivity, which is dependent on the adsorption strength of propylene. We applied a d-band center analyses of surface iron atoms and crystal orbital Hamilton population (COHP) analyses of the bonds between propylene and the surface, which demonstrates that the high selectivity of carbon-rich surfaces originates from the disruption of surface Fe ensembles caused by the spacing effect of surface carbon.

596567 Leveraging Quantum-Chemical Screening Methods to Guide the Discovery of Promising Metal–Organic Frameworks

Abstract:

One of the major advances in computational materials science over the last decade has been the ability to use high-throughput quantum-chemical calculations to accelerate the materials discovery process. Automated density functional theory (DFT) workflows in particular show great promise in identifying new material design rules and narrowing down regions of chemical space in ways not possible via traditional small-scale experimental or computational approaches. This is particularly appealing for applications involving metal–organic frameworks (MOFs), as their highly modular design leads to a nearly unlimited number of structures that could potentially be realized by combing the vast array of inorganic nodes, organic linkers, and geometric topologies.

Recently, we developed a high-throughput periodic DFT workflow to computationally investigate large numbers of MOFs in a robust and automated manner, with a specific focus on catalytic and gas separation applications involving the binding and activation of small molecule adsorbates. As one representative example, we highlight how this workflow has been used to identify a simple ligand-exchange procedure that can turn an otherwise unselective cobalt MOF into one that strongly and selectively binds O_2 over N_2, which is appealing for industrial air separation applications. Experimentally obtained adsorption isotherms confirm this finding and indicate that the identified material has one of the highest room-temperature O_2/N_2 selectivities of a MOF to date. More broadly, we demonstrate how our automated periodic DFT workflow has been used to construct the largest database of MOF quantum-chemical properties and the implications that this may have for materials design in the future. We briefly discuss our ongoing efforts to leverage this resource to predict the quantum-chemical properties of tens of thousands of MOFs in a matter of seconds via recent advances in machine learning.

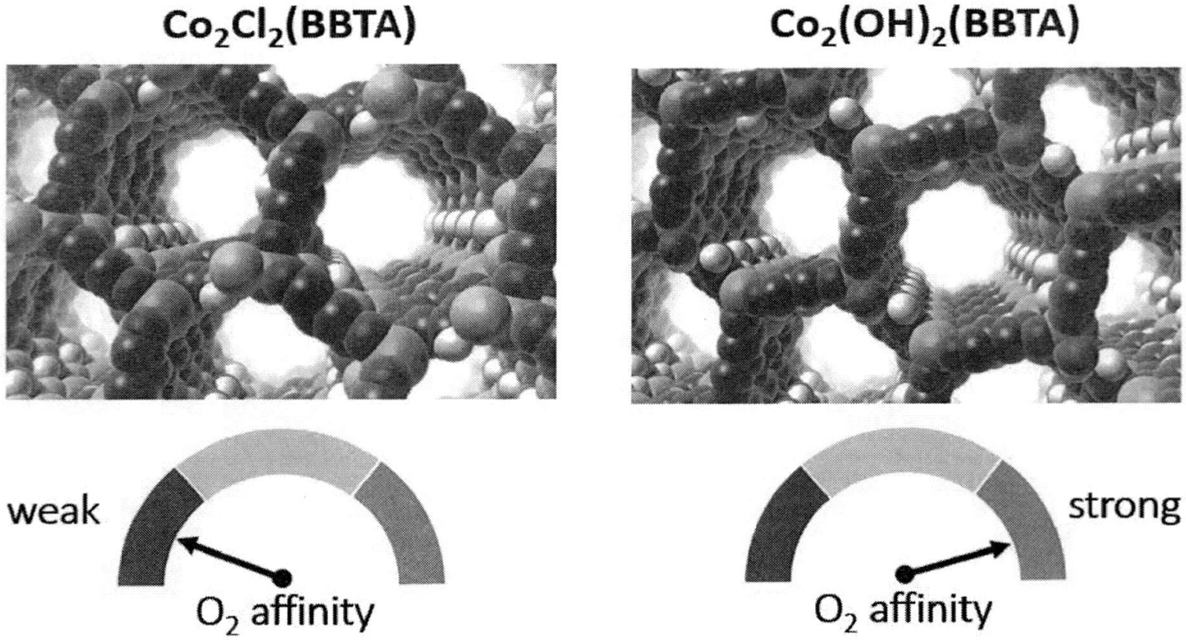

606919 Theoretical Studies on Ag-Based Homolytic Bond Dissociation Energies

Abstract:

The breaking and formation of the Ag-X (X = C, H, O, S) bonds are vital to many Ag-catalyzed/mediated reactions, such as CO_2 fixation and C-H activation. Adequate description of the bond dissociation energies (BDEs(Ag-X)) would not only help us to understand the nature of these Ag-X bonds, but also provide important instructions on rational design of Ag-based catalysts/additives. To achieve this goal, 41 diverse DFT functionals (e.g. B3LYP, B2PLYP, M06-2X, PBE, TPSSH) and two *ab initio* methods (MP2 and CCSD(T)) with different basis sets were assessed to benchmark against 6 homolytic BDEs (i.e. BDE(Ag-H), BDE(Ag-F), BDE(Ag-O), BDE(Ag-S), BDE(Ag-CH_3) and BDE(Ag-OH)) derived from experiments and/or CBS limit standards. Considering both the accuracy and computational expense, the theoretical protocol BLYP with the basis set cc-pVTZ-PP for Ag and may-cc-pVTZ for other atoms was found superior, with a precision of 1.2 kcal/mol. According to the benchmarks, the common methods B3LYP and M06-2X were found to underestimate the Ag-X BDEs while PBE overvalues them; higher-level basis sets could increase the accuracy. Further, with the aid of the calibrated theoretical methods, the ligand effect (L_nAg-X, L=H_2O, CO, CO_2 and NH_3 (n=1-5)) and the substituent effect (Ag-CH_2R, R=H, OH, SH, CH_3, Cl, etc) were systematically investigated. Note that after ligation with simple groups the Ag-X bonds became stronger compared to bare Ag-X bonds. Interestingly, for the ligands H_2O and CO, the Ag-X BDEs were lowest when n=3.

597814 Navigating Combinatorial Challenges in High-Throughput Transition Metal Complex Discovery

Abstract:

Computational high-throughput virtual screening workflows using first-principles methods such as density functional theory (DFT) have emerged as important tools for discovering new materials such as transition metal complexes (TMCs). However, even with the aid of such workflows, exhaustive design space exploration remains an intractable challenge because the space of possible complexes grows combinatorially with the number of ligands, metals, oxidation states, and ligand field symmetries considered. To address this challenge, machine learning models trained on topology-based molecular descriptors such as revised autocorrelations (RACs) have been used to rapidly and accurately assess molecular properties. RAC-trained neural networks and kernel ridge regression models are able to achieve test set errors on the scale of 1-3 kcal/mol with respect to DFT training data, in part because they are relatively sparse descriptors that average together properties of ligands in the same plane (e.g., equatorial or axial ligands in an octahedral mononuclear transition metal complex). This approximation has enabled such good accuracy even on comparatively small (ca. 100s to 1000s of points) data sets. These RAC descriptors can fail to distinguish low-symmetry complexes or other spatial isomers for which geometric information is crucial. This challenge suggests a need to better understand how molecular properties vary between ligand field symmetries that are poorly distinguished by current featurization methods. To address this, we fully enumerate the space of transition metal complexes constructed from up to two ligands from a pool of nine monodentate ligands, resulting in over 8,000 attempted DFT geometry relaxations. We quantify the extent to which properties such as adiabatic spin-splitting energy, ionization potential, and frontier orbital energetics vary between isomers within this design space and evaluate the performance of topological descriptors for predicting property variations. This analysis is then used make recommendations for the strategic exploration of broader design spaces.

598785 High-Throughput XAFS for Machine-Learning Accelerated Discovery of ALD Ternary Oxides

Abstract:

Titanium dioxide, TiO_2, is an important catalytic material with a multitude of uses in the energy sector including electrolysis, carrier extraction, air purification, and water treatment. Most relevant to work in electrocatalysis is its application as a multifunctional coating, capable of electronic tuning and corrosion stabilization. Modification of TiO_2 through the intentional introduction of transition metals or main group elements allows for control over parameters such as band position, carrier transport, interfacial electron transfer, and surface active-sites.

In this study, titanium manganese oxide ($TiMnO_x$) thin films were grown using atomic layer deposition (ALD). The addition of Mn interlayers introduces a tunable intermediate band which can be modified through varying the concentration or oxidation state of the manganese.

The amorphous films were annealed in air or in an argon atmosphere at 400°C, 500°C, and 600°C resulting in various degrees of crystallinity, a parameter which is correlated with electronic transport and stability. Such a figure of merit, as summarized in the Figure, is quantified with high-throughput acquisition and processing of synchrotron x-ray absorption data. It will also be cross validated with x-ray diffraction and TEM characterization, but only for a limited subset in a vast parameter space of ALD processing conditions.

The biggest challenge is that even though a vast parameter space can be accessed via ALD synthesis, including compositions, porosity, and atomic structures which can match up with high throughput characterizations available to us, including synchrotron x-ray absorption and x-ray diffraction, we must pick a subset of samples instead of exploring a vast parameter space. So far, many properties such as intermediate-band electronic structures can only be measured at low throughput.

We propose to explore the use of machine learning (ML) to accelerate the discovery of this new multi-functional coatings, specifically, ternary metal oxides synthesized by ALD deposition with atomic-scale structural control. ML will help mine data and establish a correlation model from the local chemical environment suggested by XANES data and suggest precise processing condition to obtain desired structures. ML is particularly powerful when our understanding of structure-property relationship is not complete.

596536 Microscopic and Machine Learning Modeling of Film Growth in Plasma Enhanced Atomic Layer Deposition

Abstract:

Plasma enhanced atomic layer deposition (PEALD) is one of the most widely adopted deposition methods used in the semiconductor industry [1]. It is chosen largely due to its superior ability to produce ultra-thin high-k dielectric films, which are needed for the further miniaturization of the microelectronic devices with the pace of the Moore's Law [2]. In contrast to the traditional thermal atomic layer deposition (ALD) method, PEALD allows for high deposition growth per cycle (GPC) under low operating temperature with the help of high energy plasma species. Despite the experimental effort in finding new precursors and plasmas, the detailed surface structures and reaction mechanism in various PEALD processes remain hard to understand because of the limitation of current in-situ monitoring techniques and the deficiency of the first-principles based analysis. There are some PEALD simulation models that attempt to reproduce film features, but are too computationally expensive to capture the surface reaction detail for an industrial scale operation [3] .

Therefore, in this work, an accurate, yet efficient DFT-based kinetic Monte Carlo (kMC) model and an associated machine learning (ML) analysis are proposed to capture the surface detail of the HfO_2 thin-film PEALD using Tetrakis-dimethylamino-Hafnium (TDMAHf) and oxygen plasma. First, Density Functional Theory (DFT) calculations are performed to obtain the key thermodynamic and kinetic parameters. Then, using those computed parameters, a detailed Kinetic Monte-Carlo (kMC) model is constructed to describe the surface reaction mechanism and to account for the structure details like steric hindrance caused by the bulky precursor. After the model is validated by experimental data, a database is generated for the cycle completion time under different operating conditions. A feedforward neural network is then constructed to characterize the relationship between partial pressure, temperature and deposition completion time.

[1] Ishikawa, K., Karahashi, K., Ichiki, T., Chang, J.P., George, S.M., Kessels, W., Lee, H.J., Tinck,S., Um, J.H., Kinoshita, K., 2017. Progress and prospects in nanoscale dry processes: How can we control atomic layer reactions? Japanese Journal of Applied Physics 56, 06HA02.

[2] Joo, J., Rossnagel, S.M., 2009. Plasma modeling of a PEALD system for the deposition of TiO2and HfO2. Journal of Korean Physical Society 54, 1048.

[3] Shirazi, M., Elliott, S.D., 2014. Atomistic kinetic Monte Carlo study of atomic layer deposition derived from density functional theory. Journal of Computational Chemistry 35, 244–259.

601109 Discovery of Self-Assembling π-Conjugated Peptides By Active Learning-Directed Coarse-Grained Molecular Simulation

Abstract:

In this work we integrate coarse-grained molecular dynamics simulation, deep representational learning, and Bayesian optimization to discover pi-conjugated peptides capable of self-assembling into biocompatible optoelectronic nanoaggregates. The pi-conjugated peptides studied in this work are triblock molecules consisting of a central aromatic core flanked by peptide wings. This class of molecules have surfaced as an extensible building block for self-assembling electronics as they have experimentally been demonstrated to form mesoscopic fibers micrometers in length and nanometers in diameter, where overlaps between pi-orbitals in these supramolecular assemblies lead to the emergence optical and electronic properties. Edisonian trial-and-error discovery of these molecules through either experiment or simulation is rendered impossible due to the combinatorial exploration in the molecular design space of pi-cores and peptide wings. We efficiently navigate the design space in search of high-performing candidates by deploying an active learning procedure which integrates three machine learning components: (i) an unsupervised deep representation learning approach to learn continuous low-dimensional embeddings of the discrete molecular design space, (ii) a supervised surrogate model using Gaussian process regression to predict molecular performance measured in simulation as a function of this embedded space, and (iii) a Bayesian optimization of the surrogate model to dictate which molecules should be evaluated next. Using this protocol, we derive a converged surrogate model for predicting molecular performance of one particular peptide family comprising tetrapeptide wings and an oligophenylenevinylene pi core after sampling only 2.3% of the design space. We identify molecules we predict to possess unprecedented self-assembly behavior and optoelectronic activity while uncovering design rules to guide the rational engineering of these molecular systems.

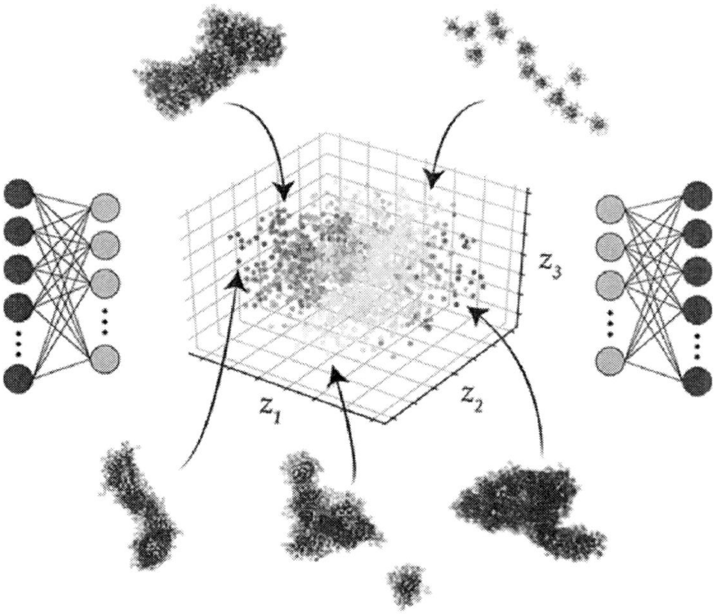

605519 Discovery of Chemical Descriptors/Features in Predicting Protein-Ligand Binding Properties Using 3D Convolutional Neural Network

Abstract:

Structure-based virtual screening techniques are some of the most successful methods for enhancing various processes of interest such as artificial dialysis, drug delivery, purification of specialty chemicals, sensing and detection, and waste-water treatment. With the growing needs for advanced materials that facilitate low energy separation, capture or delivery of molecules with high specificity, there has been a significant effort by the researchers to apply advanced Machine Learning (ML) models to estimate material properties and design novel materials. One such application is to predict protein-ligand binding pockets and/or affinities using 3D Convolutional Neural Networks (CNN). Here, the 3D protein/ligand structure is treated as a 3D image and pharmacophore properties are used as the features/channels for the CNN. This approach suffers from two basic limitations: the features generated should be rotationally and translationally invariant and only the relevant features should be used to train the network. To eliminate these limitations, in this study, we use an *Euclidean Neural Network,*[1] which is a CNN with specially designed convolution kernel that transforms the input molecular features such that they are rotationally and translationally invariant and therefore leading to more robust learning of the network. At the same time, we aim to incorporate feature extraction directly into the ML model. In such an approach, the CNN described above will become the first component of the model, which is then trained together with a predictive component to extract features that are useful in solving a specific task such as predicting protein-ligand binding properties. With this learn-test-design approach, it is possible to discover specific features/descriptors and identify relationships relevant to the specific objective, which will lead to the design of novel materials for particular applications.

(1) Thomas, N.; Smidt, T.; Kearnes, S.; Yang, L.; Li, L.; Kohlhoff, K.; Riley, P. Tensor Field Networks: Rotation- and Translation-Equivariant Neural Networks for 3D Point Clouds. **2018**.

606274 Hierarchical Assembly of Peptoids into Complex Nanostructures

Abstract:

Peptoids are complex and diverse oligomeric structures which have been explored for a number of applications including drug molecules, surfactants of catalysts.[1-2] In contrast to their amino-acid composed peptide counterparts they are highly flexible and diverse, spanning a large design space of chemical and structural functionality through the careful tuning of their side chains. One novel example of this is the N-substituted α-chiral, aromatic side chains oligomers which can produce stable helical structures in both aqueous and non-aqueous solutions.[3] These oligomers have been shown to have the capacity to assemble into more complex hierarchical assemblies like microspheres, nanosheets, and dynamic single-walled nanotube structures.[4] However, little is known about the mechanisms of these assemblies. Molecular dynamics provide an opportunity to understand this dynamic behavior at an atomistic-level of detail. Here we present our extension of the MFTOID peptoid forcefield[5] to investigate the mechanisms that give rise to these unique complex assemblies.

[1] Dohm MT, Kapoor R, Barron AE. Peptoids: Bio-Inspired Polymers as Potential Pharmaceuticals. *Curr. Pharm.* **2011,** 47 (17), 2732

[2] Maayan G, Ward MD, Kirshenbaum K. Folded biomimetic oligomers for enantioselective catalysis. *Proc Natl Acad Sci* **2019**, 106(33):13679-84

[3] Wu C, Sanborn S, Huang K, Zuckermann RN, Barron AE. Peptoid Oligomers with α-Chiral, Aromatic Side Chains: Sequence Requirements for the Formation of Stable Peptoid Helices. *J. Am. Chem. Soc.* **2001**, 123, 28, 6778-6784

[4] Weiser LJ, Santiso EE, Molecular modeling studies of peptoid polymers. *AIMS Materials Science*, **2017** 4(5): 1029-1051.

[5] Mirijanian DT, Mannige RV, Zuckermann RN, et al. Development and use of an atomistic CHARMM-based forcefield for peptoid simulation. *J Comput Chem.* **2014.** 35: 360–370

599928 Simulation-Guided Tuning of Sequenced Peptoids Towards Achieving Shape-Controlled Au Nanocrystal Growth

Abstract:

Metal nanocrystals have remarkably advanced cutting-edge technologies owing to their unique plasmonic, electronic and catalytic properties, which are highly dependent on their shapes and sizes. A broad spectrum of nanostructures has been synthesized in solution with chemical additives. These chemical additives often show facet selectivity and can impart shape-controlled growth thermodynamically, kinetically, or both. The identification of these chemical additives, however, is still based on a "trial and error" process experimentally and only a small library of effective structure-directing chemical additives has been found so far. Thus, a less expensive and theory-based selection of such chemical additives is urgently in demand.

Peptoids are synthetic polymers that have similar backbone structure as peptides, with side chains connected to the backbone nitrogen atoms instead of the -carbons. Some peptoids have been shown to effectively assist shape-controlled metal nanocrystal formation. Because peptoid sides chains come in a variety of chemical structures and their sequence is tunable, they can serve as excellent templates to be engineered towards chemical additives for shape-controlled nanocrystal growth. We use molecular dynamics simulations and enhanced sampling methods to obtain adsorption free energies of several side chain analogs and understand their adsorption mechanisms. Based on theoretical findings from previous studies on how facet selectivity can influence nanocrystal growth, we program peptoid sequences to achieve a facet selectivity that may bring a desired outcome. We verify the energetic facet selectivity of peptoids by calculating their adsorption free energies using Parallel Bias metadynamics. These peptoids have effectively guided the experiments to realize desired nanostructures.

605481 Generating Potential Energy Landscapes for Membrane Proteins Using High-Throughput Simulations

Abstract:

Integral membrane proteins are ubiquitous in biological cellular and subcellular membranes. Despite their significance to cell function, isolation of membrane proteins from their hydrophobic lipid environment and further characterization remains a challenge. Several computational approaches, such as docking or self-assembly simulations, have been used to obtain insights into membrane proteins; however, the promise of these approaches has been limited due to the computational cost. Here we present a new method called Protein AssociatioN Energy Landscape (PANEL) that provides an extensive and converged data set for all possible conformations of membrane protein associations using a combination of stochastic sampling and equilibration simulations. The PANEL method generates a potential energy landscape describing the thermodynamic favorability of protein-protein association states, and a frequency landscape representing their kinetic ease of formation. Based on this comprehensive dataset, we constructed a systematic approach to rank the dimer orientation taking both relative stability of the protein dimers and their formation frequency into consideration. We mapped this data to the features on the landscape—Funnels—deep thermodynamically favored states that may be kinetically less favored, and—Saucers—shallow, thermodynamically less stable states, yet wide landscapes that are kinetically favored. Using the Funnel and Saucer characterization of the landscape, we demonstrate an efficient way of identifying significant protein-protein interactions from a diverse set of protein families. The PANEL method is applicable to all membrane proteins and is freely distributed for use by the research community.

602648 Introduction of Alchemical Variables in Metadynamics to Enhance Configurational Sampling

Abstract:

Over the past decades, various advanced sampling methods have been developed to address the challenge of comprehensive sampling of the configurational space of systems with a rough free energy surface. As adopted by several well-known algorithms, one common idea of accelerating the sampling is to modify the underlying free energy surface. For example, expanded ensemble utilizes the Wang-Landau algorithm to calculate the biasing weight for each alchemical state based on how frequently the state is visited. While this method circumvents the need of defining a set of collective variables, the system may get stuck in the alchemical space. To alleviate this issue, here we propose to add alchemical variables as an allowed collective variable in metadynamics. With this add-on, metadynamics is no longer restricted to only sampling the configurational collective variables, but is able to sample the alchemical direction as well. In this study, we validate our implementation of alchemical metadynamics in PLUMED with test systems/alchemical processes with varying complexities and dimensions of collective variable space, from disappearing a lactic acid molecule from water, to the binding and unbinding event of CB8-G3 host-guest binding complex. One-dimensional alchemical metadynamics correctly recovers the free energy difference between the coupled state and the uncoupled state. In addition, multi-dimensional alchemical metadynamics has the potential to further facilitate the sampling by introducing configurational collective variables to circumvent barriers. The method can trivially be combined with other metadynamics-based algorithms implemented in PLUMED.

611727 Advancing Rational Design of Peptide Self-Assembly at Surfaces

Abstract:

Rational control of self-assembly of peptides in interfacial environments is crucial for creating tailored bionanomaterials and to realize their biomedical and commercial applications. [1-3] Recent studies demonstrated promising approaches using designed mutations to control structure and assembly of peptides at graphene and other nanomaterials. [4-7] In designing mutations, both location of the mutation and mutant residues are two critical and possibly dependent choices. The former is to make a targeted structural change and the latter is for realizing it. Knowledge on the free energy of adsorption (ΔA_{ads}) of amino acids or interaction energies of individual residues on surfaces have been used as a guide for selecting mutant residues with some success. [4-7] To further in this selection process and advance designed mutations approach, knowledge on dependence of the ΔA_{ads} of a residue on other residues is also required. Our goal is to work towards it and we start by developing insights into the dependence of ΔA_{ads} and structure of a residue on their neighboring residues. For this, we use graphene as our model surface and 36 model tripeptides with motif LNR-CR-Gly, where LNR and CR are variable left neighboring and central residues, respectively.

We consider a combination of strongly adsorbing (aromatic, Arg) and weakly adsorbing (Val, Leu, Ser, Thr) amino acids using ΔA_{ads} of amino acids estimated in a prior study. [8] Our results indicate that ΔA_{ads} of tripeptide is not the sum of ΔA_{ads} of each residue. However, the contributions from the strongly adsorbing amino acids was dominant suggesting that such residues could be used for strengthening peptide-graphene interactions irrespective of their neighboring residues. On the other hand, our structural analysis revealed that the dihedral angles of LNR and CR are more dependent in the adsorbed state than in bulk state. Together with ΔA_{ads} trends, this implies different backbone structures of a given CR can be realized for a similar ΔA_{ads} depending on LNR. Therefore, care has to be taken in designing mutations by incorporating context effects to engineer peptide structure at graphene surface. We will present these results and also discuss approaches using machine learning in designing mutations to control peptide assembly at surfaces.

References:

1. Tadi KK, Narayanan TN, Arepalli S, Banerjee K, Viswanathan S, Liepmann D, Ajayan PM, Renugopalakrishnan V. Engineered 2D nanomaterials–protein interfaces for efficient sensors. *Mater. Res.* 2015 Dec;30(23):3565-74.
2. Wang L, Zhang Y, Wu A, Wei G. Designed graphene-peptide nanocomposites for biosensor applications: A review. *Analytica Chimica Acta.* 2017 Sep 8;985:24-40.
3. Walsh TR, Knecht MR. Biomolecular Material Recognition in Two Dimensions: Peptide Binding to Graphene, h-BN, and MoS2 Nanosheets as Unique Bioconjugates. *Bioconjugate Chem.* 2019 Oct 8;30(11):2727-50.
4. No YH, Kim NH, Gnapareddy B, Choi B, Kim YT, Dugasani SR, Lee OS, Kim KH, Ko YS, Lee S, Lee SW. Nature-inspired construction of two-dimensionally self-assembled peptide on pristine graphene. *Phys. Chem. Lett.* 2017 Aug 17;8(16):3734-9.
5. Zou X, Wei S, Jasensky J, Xiao M, Wang Q, Brooks III CL, Chen Z. Molecular interactions between graphene and biological molecules. *Am. Chem. Soc.* 2017 Jan 16;139(5):1928-36.

   Orientation on Graphene. *Am. Chem. Soc.* 2019 Nov 27;141(51):20335-43.
7. Hayamizu Y, So CR, Dag S, Page TS, Starkebaum D, Sarikaya M. Bioelectronic interfaces by spontaneously organized peptides on 2D atomic single layer materials. *Scientific Reports.* 2016 Sep 22;6:33778.
8. Dasetty S and Sarupria S., "Adsorption of Amino Acids on Graphene: Assessment of Current Force Fields", *Soft Matter*, 15, 2359-2372

605487 Transition Path Sampling Simulations of Base Flipping in RNA

Abstract:

Base flipping is a key biophysical event involved in recognition of various ligands by nucleic acids. Therefore, the dynamics of spontaneous flipping in/out of a mismatched base in nucleic acids has a significant implication in modifications in DNA/RNA by enzymes. While the mechanisms of base flipping in nucleic acids have been probed using experimental and computational techniques previously, our understanding of molecular scale details of this biophysical event remains limited. Specifically, it remains unclear how neighboring nucleobases, solvent molecules, and ions reorganize to trigger a base flipping event. In this work, we applied transition path sampling methods to overcome limitations of classical molecular dynamics (MD) simulations for studying rare biophysical events and obtained 1000 transition trajectories to describe the base flipping process in a double-stranded ribonucleic acid (RNA) molecule. These trajectories facilitated the selection of key collective variables (CVs) that can distinguish between the inward/outward configurations and provide an optimal reaction coordinate when combined, as tested by applying the likelihood maximization and Bayesian information criterion. The free energy surface and the transition dynamics were then projected along the optimal reaction coordinate, and the rate constant for the base-flipping event was determined.

609721 Mechanism of Inorganic Phosphate Release from Actin Subunits

Abstract:

The diverse functionalities of cells such as their ability to move, divide, and respond to environmental perturbations are governed by actin filaments. The nucleotide bound to actin subunits in the filament undergoes a series of reactions called ATP hydrolysis and inorganic phosphate (Pi) release. It is well known that the rate of ATP hydrolysis reaction is four orders of magnitude faster in actin filaments compared to monomeric actin (g-actin) and a lot of studies have focused on unraveling the key steps driving this reaction. However, the Pi release reaction is not well studied and a critical understanding of the structural changes favoring this reaction is still lacking. In this regard, here, we investigate the mechanism of Pi release reaction and correlate it with actin subunit conformational changes. Specifically, we adopt Transition Tempered Metadynamics (TTMetaD), an enhanced sampling simulation method, to efficiently sample the underlying free-energy landscape and identify the minimum free energy path for the release of inorganic phosphate. Several actin-binding proteins preferentially interact to actin subunits with a specific nucleotide state and regulate the formation of various types of higher-order actin filament structures such as crosslinked networks, bundled filaments, branched networks, and so on. A comprehensive knowledge of the mechanism of the phosphate release reaction would not only help in understanding the binding of these regulatory proteins but also in the development of rigorous bottom-up coarse-grained models, which can be used to simulate actin filament networks in an efficient manner.

596293 Machine Learning and Computational Simulation Aided Design of Antibacterial Oligothioetheramides

Abstract:

Oligothioethermamides are novel molecules developed to combat the rising crisis of antibiotic resistance. Their antibacterial properties present similarly as antibacterial peptides, and thus they have similar factors that affect their activity and toxicity, namely the backbone hydrophobicity and cationic charge. Our atomistic simulations investigating into two isomeric oligothioetheramides, with the same cationic charge and hydrophobicity have shown that the binding of these oligomers to the surface of lipid bilayers could be a suitable candidate for an objective function that could then be used to aide in their design. Oligothioethermaides have the added advantage of allowing greater control over the backbone sequence, and thus pose a large combinatorial problem that involves exploring the sequence space occupied by the backbone sequence of these oligomers. We have performed coarse-grained simulations in tandem with applying Bayesian optimization to optimize the activity of these molecules and provide suitable candidates for experimentalists. We have also used Support Vector Machines to train on data available in databases for antibacterial peptides. By developing similar data for the combinatorial space of oligothioethermides by mapping and extracting similar parameters, we applied the trained machine on this unexplored space to categorize the oligomers. We thus compared the candidates provided by these two machine learning techniques to measure the performance of our objective function.

598435 Monomer-Based Kinetic Monte Carlo Simulation for Multifunctional Polymerization

Abstract:

Complex polymer networks are abundant in both nature and art, as their crosslinked structure is exploited for its insolubility and stability. Multifunctional monomers comprised of multiple, diverse functional sites of interest here polymerize into these complex polymer networks through a free-radical mechanism. The heterogeneity of the resulting polymer and products increases via oxidation and scission reactions. Efforts have been made to capture the diverse product distribution, but the nature of radical polymerization leads to a combinatorial explosion of species and limits tractability of solving the time evolution of concentrations with a large system of ordinary differential equations in a deterministic method.[1]

This research utilizes a two-level polymer framework and stochastic kinetic Monte Carlo (kMC) simulation to address the challenges faced with modeling complex polymer systems. Polymers are graph-like: monomer nodes are crosslinked by covalent bond edges. Further, each monomer type is a unique graph with individual atoms and covalent bonds. Chemical reactions transform graphs with specific node patterns or functional groups by making and breaking edges. In literature, Orlova *et al.* tracked chemistry at the monomer level, generating unique monomer units, but they did not maintain explicit crosslinked species and instead used random graph theory to reconstruct polymer distribution at discrete time points.[2] Kurdikar *et al.* also used a monomer-based model for multifunctional polymerization, tracking the evolution of unique polymers by the collection of monomer units they contained, but not the polymer topology of how monomers are connected.[3]

This work leverages both monomer approaches while retaining crosslink information. Kinetically informed stochastic selection of reaction type, then species, then functional group are identified and reacted to the product. The graphs of polymer species are stored as an adjacency list of monomer IDs and their respective connections and bond types, such as alkyl, ether, or peroxyl crosslinks. Graph isomorphism allows for species uniqueness checks.

Using an on-the-fly, rule-based kinetic Monte Carlo scheme in which possible reaction events are generated dynamically allows kinetically significant reaction pathways to be tallied in real time, preventing artificial truncations of species size dictated by a pre-populated, finite list of reactions that can lead to a skewed product distribution. The advantage of the dynamic kMC approach is demonstrated using a toy model artificially truncated to dimers compared to one generated using on-the-fly, rule-based product evolution based on free-radical polymerization with representative rate constants involving 15 monomer bases.

Developing a computational tool that captures polymer topology and maintains atomic-level detail for complex, multifunctional polymers allows for deeper exploration of the polymer distributions, their reactivity, and their sensitivity to process parameters in a computationally feasible manner.

[1] Oakley, L. H., Casadio, F., Shull, K. R., & Broadbelt, L. J. (2015). Microkinetic modeling of the autoxidative curing of an alkyd and oil-based paint model system. *Applied Physics A*,

[2] Orlova, Y., Kryven, I., & Iedema, P. D. (2018). Automated reaction generation for polymer networks. *Computers & Chemical Engineering*, *112*, 37–47.

[3] Kurdikar, D. L., Somvarsky, J., Dusek, K., & Peppas, N. A. (1995). Development and Evaluation of a Monte Carlo Technique for the Simulation of Multifunctional Polymerizations. *Macromolecules*, *28*(17), 5910–5920.

606557 Utilization of Molecular Dynamics to Predict Glass Transition Temperatures of Imidazolium-Based Ionic Liquids and Their Mixtures

Abstract:

The inherent characteristics of ionic liquids make them popular candidates as electrolyte solutions, and their properties are also easily tunable via selection of constituent ions for desired applications. When designing ionic liquid-based electrolytes, the glass transition temperature of the solution is taken into consideration as it determines the operating temperature window while also correlates with key transport properties such as viscosity and conductivity. Identifying and predicting the glass transition temperature of ionic liquids and their mixtures can aid in expanding the design, optimization, and implementation of their use. In this study, we report a method to predict the glass transition temperature of ionic liquids, 1-propyl-3-methylimidazolium iodide and 1-butyl-3-methylimidazolium iodide, and their mixtures with water and gamma-butyrolactone, respectively, by utilizing molecular dynamics (MD) simulations. Annealing simulations over a temperature range of 500K to 100K are performed to effectively capture the glass transition phase. The potential energy of each system is monitored to reveal variations in behavior that could indicate a glass transition temperature. Initial results suggest that parameters such as simulation box size and cooling rate may have a significant impact on the capability to successfully predict the glass transition temperature. To understand underlying molecular interactions, radial distribution functions are computed at varying temperatures. Other properties including self-diffusion coefficients and densities are also calculated to check for observable changes near the glass transition temperature. The simulation results are complemented by experimental characterization methods, including dynamic differential calorimetry (DSC) and environmentally controlled rheology, to validate and further improve this reported computational method in predicting the glass transition temperature of ionic liquids and their mixtures with cosolvents.

605685 Machine Learning the Fundamental Tradeoffs between Conductivity and Voltage Stability in Solid State Electrolytes

Abstract:

All solid state batteries provide many safety advantages over traditional lithium-ion batteries by replacing the combustible organic liquid electrolyte with a ceramic solid-state electrolyte. However, reported superionic conductors with conductivities approaching that of liquid electrolytes are unstable in contact with a lithium anode leading to increased internal cell resistance, and poor cyclability. Conversely, compounds stable at the anode or cathode interfaces do not exhibit useful bulk ionic conductivities. Although *ab initio* methods exist to study each ionic conductivity and voltage stability range, there is no established theory to connect these two properties. Here, we leverage machine learning (ML) to investigate the role of crystal structure in the tradeoff between voltage stability and ionic conductivity. To this end, we trained a partial least squares (PLS) machine learning algorithm using the valence electronic density both in real and in reciprocal space as descriptors of 150 known solid-state electrolytes along with their corresponding ionic conductivity, anodic voltage limit, and cathodic voltage limit. The trained model has an 80% prediction accuracy and suggests that within the search space of crystal structures, the voltage stability and ionic conductivity are inherently inversely correlated (Figure 1a). The real space electron density descriptor was very sensitive to small variations in the signal resulting in estimated uncertainties of 3 orders of magnitude in ionic conductivity. Although the reciprocal space electron density was less with an uncertainty of 1 order of magnitude (approaching that of *ab initio* molecular dynamics), the latent variables were more difficult to interpret. (Figure 1c). The PLS model successfully identifies BCC anion substructure (Figure 1b) and channels as effective descriptors, which is in good agreement with prior work.[1] Using this model, we screened through a database of ca. 14,000 materials and identified five new promising solid state electrolyte candidates to have conductivities greater than 16 mS/cm. The model predictions were subsequently verified with *ab initio* molecular dynamics simulations. The proposed ML model and electron density descriptor may be used in future studies to elucidate structure-property relationships for other applications with high accuracy and without sacrificing interpretability.

References:

[1] Wang, Y. *et al.* Design principles for solid-state lithium superionic conductors. *Nat. Mater.* **14,** 1026–1031 (2015).

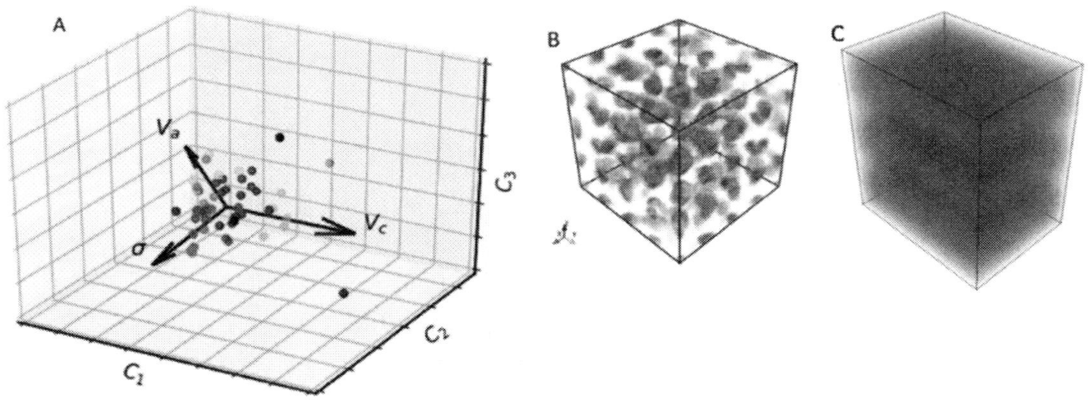

Figure 1: (a) Latent variable space determined by the PLS model used to predict properties. First latent variable in (B) real space, and (C) reciprocal space. The real space variable highlights the BCC anion substructure identified by the PLS model, and the reciprocal space electron density could signify 2D materials or characteristic distances between lithium sites.

607996 Computational Modeling to Understand the Spatiotemporal Cholinergic Modulation of Hippocampal Synaptic Plasticity

Abstract:

In septohippocampal cultures, stimulating the Schaffer collateral to CA1 pyramidal neuron synapse while optogenetically stimulating the stratum oriens (releasing acetylcholine) has been shown to induce long term potentiation, short term potentiation, and short-term depression mediated through either muscarinic or nicotinic receptors. The type of induced plasticity and the selectivity of receptors is dependent on the relative timing of cholinergic input and Schaffer collateral stimulation. We hypothesize that receptor desensitization is a crucial determinant of the observed receptor selectivity and, therefore, the induction of the specific form of plasticity. The desensitization time of muscarinic and nicotinic receptors have been shown to differ by several orders of magnitude. In this work, we develop a physiologically relevant computational modeling framework to investigate how the temporal activation of the Shaffer collateral-CA1 pyramidal neuron synapse with cholinergic inputs impacts the specific receptors type deactivation and leads to the induction of the particular form of synaptic plasticity mediated through different types of cholinergic receptors. We find that the desensitization time impacts the induction of a particular form of plasticity by altering the intracellular calcium transients. Our findings elucidate the effects of neuromodulation on the synaptic plasticity through receptor desensitization and describe how temporal activation of cholinergic inputs modulate the synaptic connections in the hippocampus.

600013 Towards Predicting Kinetics in Systems of Coupled Reactions with Deep Neural Networks

Abstract:

The computational cost of modeling kinetics atomistically grows with the system's size.[1,2] Therefore, this approach becomes unfeasible for large networks of reactions and few efforts have been made in this direction.[3] Recently there has been interest in using machine learning models to predict kinetics.[4] Indeed, trained deep neural networks (DNNs) have the potential to alleviate the computational cost of atomistic simulations[5] by making predictions based on easily computable static properties such as optimized reactant and product geometries and ground state energies.[6] In this context, we aim to understand whether machine-learned rate constants can accurately describe the populations of chemical species in a network of reactions. Hence, we trained DNN models using existing databases of gas-phase kinetic data to predict reaction rate constants. Optimal DNN hyperparameters were identified by exploring the structure of the model, and input features were selected according to their importance in predicting the rate. We then looked at a small set of coupled gas-phase reactions as a testing set and used our trained DNN to predict the rate constants. The accuracy of the DNN at the level of network populations was evaluated by comparing the exact model with the one predicted by the DNN.

1. Ju, L. P., Han, K. L. & Zhang, J. Z. H. Global dynamics and transition state theories: Comparative study of reaction rate constants for gas-phase chemical reactions. *Journal of Computational Chemistry* (2009) .

2. Sumathi, R. & Green, W. H. A priori rate constants for kinetic modeling. *Theoretical Chemistry Accounts* (2002) .

3. Wang, L. P. *et al.* Discovering chemistry with an ab initio nanoreactor. *Nat. Chem.* (2014) .

4. Gao, H. *et al.* Using Machine Learning to Predict Suitable Conditions for Organic Reactions. *ACS Cent. Sci.* **4**, 1465–1476 (2018).

5. Valleau, S. & Martínez, T. J. Reaction Dynamics of Cyanohydrins with Hydrosulfide in Water. *J. Phys. Chem. A* (2019) .

6. Collins, C. R., Gordon, G. J., Von Lilienfeld, O. A. & Yaron, D. J. Constant size descriptors for accurate machine learning models of molecular properties. *J. Chem. Phys.* **148**, (2018).

607727 Polyatomic Ion Intercalation into Thiophene-Based Oligomers: Thermodynamics and Impacts on Inter-Chain Charge Transfer

Abstract:

Conjugated polymers that have been electrochemically doped via ion injection are at the cutting edge of solid-state electrolytes for organic photovoltaics. They benefit from a number of advantages, including flexibility that permits spin coating onto substrates, and a diversity of backbone moieties and side chain functionalities. These properties permit some degree of ion permeation into the semiconductor, which bestows an advantage over inorganic materials. The performance of resulting devices is sensitive to the properties of both the electrolyte and polymer, particularly electrochemical oxidation tendencies of the polymer and the compensatory ability of counterions that contribute to bulk charge mobility. Ion uptake is coupled to the polymer's local structure in a complicated way and is nonmonotonic with ion size, an interplay that must be well-understood to successfully design novel technologies. Addressing ion mobility in bulk polymer, empirical simulations were conducted with atomistic oligomers of side-chain functionalized thiophene (a benchmark organic semiconductor) in contact with aqueous solutions of polyatomic ions. Snapshots from the equilibrium distribution provide coordinates for quantum mechanical calculations, followed by Marcus Theory to study the energetics of electron hopping in proximity to a counterion. Polyatomic counterions that include hydrophobic components, like bistriflimide ion, experimentally show a greater tendency to permeate the low-dielectric polymer than exposed charges, despite the extra volume taken up by the appendage. Empty space is observed to be distributed inhomogeneously within the polymer melt, prompting our hypothesis that there are local voids sufficient for large ions to move unencumbered over short distances. As electron holes are transferred between adjacent chains, the counterion reorientation response time may not be sensitive to size. These insights will aid in the inference of other polymer-ion pairings that will produce efficient photovoltaics.

608559 General-Purpose Coarse-Grained Toughened Thermoset Model for Aerospace Composites: Workflows, Parameterization, and Validation

Abstract:

The efficient manufacturing of composite aerospace materials can be improved with insights from molecular simulations when sufficiently large volumes and timescales can be accessed. Predicting the structure of aerospace composites is especially challenging because of the high computational cost of relaxation (scales cubically with polymer chain length), morphological features spanning dozens to millions of repeat units, and the nonequilibrium dynamics of reacting epoxy thermosets. Coarse-grained simulations can overcome these barriers by using fewer simulation elements to represent the same number of atoms and simplified reaction mechanics which both accelerate sampling dynamics. In this work we deploy coarse-grained simulations of reacting epoxy thermosets and validate them against two common chemistries with the longer term goal of improving the fuel efficiency of airplanes of flight and aerospace manufacturing.

We use a single site coarse-grain model to represent each epoxy, amine, and toughner repeat unit. We implement a dynamic bonding plugin with HOOMD-Blue to simulate the cross-linking of the epoxy and amine species. We present workflows for parameterizing models which include deriving Lennard-Jones parameters from cohesive energy data and empirically deriving the kinetic parameters of our model from experimental curing data.

We validate our model using data from curing experiments of the amine monomer 44DDS with both DGEBA and TGDDM epoxy in stoichiometric ratios, both toughened with PES. We compare predicted morphologies against experimental scattering and microscopy data, as well as glass transition data. We find our model correctly predicts glass transition temperatures as a function of degree of cure, the phase behavior, and toughener domain sizes in 73 nm x 73 nm x 73nm volumes. We also demonstrate our model's sensitivity to cure profile which enables studies of how processing conditions affect cure time, mechanical, and thermodynamic properties. This work demonstrates tooling to better understand the relationship between structure, processing, and properties in toughened epoxy thermosets for improving the efficiency of composite aerospace part manufacturing.

601126 Molecular Weight Transferable Electronic Structure Prediction for Coarse-Grained Polymers

Abstract:

Soft materials modeling for non-crystalline systems involves the integration of multiscale classical and quantum mechanical techniques. Recently, we introduced a methodology known as electronic coarse-graining that relies upon techniques from supervised machine learning to make accurate quantum-chemical predictions directly from coarse-grained degrees of freedom, thereby rapidly accelerating quantum-mechanical predictions for soft materials. In this talk, I will outline the extension of this methodology to polymers, specifically addressing the problem of the molecular weight transferability of electronic coarse-graining predictions. Focus will be paid to the integration of machine learning techniques with phenomenological quantum-mechanical Hamiltonians, enhancing transferability and interpretability of electronic structure simulations at a coarse-grained resolution.

599601 Development of Coarse-Grained Models to Capture Specific and Directional Interactions in Macromolecular Materials

Abstract:

Specific and directional interactions like hydrogen bonding offer an attractive route to precisely tune structure and morphology within macromolecular materials. Computational studies on macromolecules with directional interactions have been largely limited to atomistic simulations with explicit representation of hydrogen bonding atoms making them expensive for simulation of directional interactions in polymers. On the other hand, simulations with coarse-grained (CG) models can capture the large length and time scales associated with polymer chains but cannot capture the effects arising due to the specificity and directionality of interactions without using complex anisotropic interaction potentials to model directional interactions.

In this poster we will present our recent work on development of a coarse-grained model that captures short length and time scales of directional and specific interactions as well as the large length and time scales of polymer chains. We will describe pertinent details of the model development that allow us to produce effectively anisotropic attractive interactions while still using isotropic interaction potentials. We will also demonstrate the use of this CG model in molecular dynamics (MD) simulations and Polymer Reference Interaction Site Model (PRISM) theory calculations to capture effects due to directional interactions in polymer nanocomposites and polymer blends.

606603 Multiscale Computational Study of Continuous Manufacturing of Polymeric Micelle Based Nanocarriers

Abstract:

Purpose:

Typical micelle-based drug delivery carriers are formed from amphiphilic block copolymers which self-assemble into core-shell nanoarchitectures with a hydrophobic core being able to encapsulate drug molecules and a biocompatible hydrophilic shell. An innovative turbulent co-flow platform of water and ethanol (with drug and polymers) was previously developed to continuously process drug loaded micelle nanocarriers. Polymer aggregation is the result of the intermolecular forces among molecules and along with the jet flow characteristics impact to the formation of micelles. However, the underlying mechanism and the detailed effects from material attributes are only partially understood. In order to investigate the underlying mechanism and quantitative estimation of the effect of material attributes and process parameters on the quality of the polymeric micelle formulation, we implemented a multiscale computational approach to study micelle formation using a coaxial turbulent jet flow.

Methods:

All-atom (AA), coarse-grained molecular dynamics (CG-MD), as well as computational fluid dynamics (CFD) simulations have been conducted to not only reveal the effects to material attributes and processing parameters during the micelle formation, but also as parametric case studies for this process. The initial conditions applied into the simulations were based on polymeric micelle experiments in which PEG-PLA (2kD-1.7kD) was used as model drug carrier. The CFD simulation was implemented using Large Eddy Simulation (LES) model in COMSOL Multiphysics incorporating energy equation and high-resolution mesh in mixing area. The MD simulation trajectories and their analysis were carried out using GROMACS package. We applied both CHARMM and MARTINI force-fields for all-atom as well as coarse-grained simulations, respectively. The initial coordinates of models and their force field parameters were generated by CHARMM software and MARTINI forward mapping approach. The steepest decent algorithm was used in energy minimization with a 20 fs time step followed by 10 ns equilibration step in isothermal−isochoric NVT ensemble at temperature of 300 K, then the production runs were performed beyond 1 μs in the NPT ensemble using the Nosé−Hoover thermostat and the Parrinello−Rahman barostat at pressure of 1 atm with constant temperatures at every 20 fs time steps. The simulations were carried out using periodic boundary conditions. Computations were performed in High Performance Center Supercomputer Cluster at the University of Connecticut.

Results:

The self-assembled process of polymeric micelle was studied in the system with and without drug molecules, and in each case, block copolymer was randomly packed in the simulation box with explicit water and ethanol for all simulations. With optimized CG-MD forward mapping strategy and force field parameters incorporating MARTINI standard beads and recent developed S-beads, we observed that polymeric micelles formed successfully and fell into the experimental size range. The simulation of micelles formation and the size distribution the micelles indicate that CHARMM, MARTINI force fields (FFs) can effectively capture the experiment behavior and results . By adjusting Smagorinsky coefficient (Cs) of LES model companied with energy and mass transfer equations, CFD simulations successfully modeled flow patterns and formation temperatures of co-axial turbulent jet flow from experiments, as well as successful comparison and verification. Our CG-MD and CFD simulations are in good agreement with experimental results.

Conclusions:

In this present work, the model predictions provide an understanding of the impact of drug-polymer interactions, drug-polymer ratio, initial organic phase polymer concentration, as well as maintaining the agreement between the computational predictions and experimental findings highlight the power of the multiscale approach to be able to cover the long and short length scale predictions. Moreover, the details of the formation process was revealed in micro, macro, and meso scales by connecting discrete and continuum computational modeling work which are found to be complementary to one another. The multiscale approach used in this work can be effectively utilized as a powerful tool in the process of discovery, development, and optimization of new drug delivery systems in the co-flow continuous processing.

Acknowledgements: FDA Grant# 1U01FD005773-01.

Disclaimer: This article reflects the views of the authors and should not be construed to represent FDA's views or policies.

606796 Full-Atomistic Entangled Polymer Structure Generation Using Reverse-Mapping from Breads-Spring Models.

Abstract:

Polymer materials are now widely used in our society. Weight reduction, flexibility, and toughness of the materials are the important features. Molecular simulation is a useful tool to examine materials properties from the microscopic viewpoint. Polymer entanglements are often discussed in studies of rheological feature of polymers using molecular simulation. The entangled chains also affect to the materials fracture. There is a study which claims that the entanglements and characteristic ratio are key factors determining the material ductility [1]. Hence it is important to generate polymer structures which reproduce experimental entanglement and chain stiffness characters in order to clarify materials properties using molecular simulation. Here we demonstrate full-atomistic entangled polymer structure generation using reverse-mapping technique from the beads-spring models. In the previous studies [2,3], we reported polymer structure generation scheme for amorphous bulk models and for adsorbed chains on a solid surface based on a mapping relation proposed by C. Svaneborg et al. [4]. The mechanical properties of the generated models in a deformation simulation will be discussed in this presentation.

[1] S. Wu, "Chain structure and entanglements", J. Polymer. Sci. B Polym. Phys. 27, 723 (1989)

[2] H. Nitta and T. Ozawa, "Efficient Generation of Polymer Amorphous Structure By Reverse-Mapping from Beads-Spring to Full-Atomistic Model", AIChE 2018 Annual Meeting

[3] H. Nitta and T. Ozawa, "Efficient Generation of Full-Atomistic Polymer Amorphous Structure Adsorbed onto Solid Surfaces by Reverse-Mapping technique", AIChE 2019 Annual Meeting

[4] C. Svaneborg et al., "Kremer-Grest models for universal properties of specific common polymer species", arXiv:1606.05008

609685 All-Atom Molecular Dynamics Study for O_2 Permeation in a Nafion Polyelectrolyte Membrane

Abstract:

Nafion, a perfluorosulfonic acid polymer, is the most widely studied polymeric electrolyte for Polymer electrolyte fuel cell (PEFC) applications due to its good mechanical properties, chemical stability, and high ionic conductivity.[1] In the cathode catalyst layer of PEFC, an ionomer with which the catalyst is covered is very important on the point of transferring protons to the catalytic surface on the cathode side. On the other hand, it is said that ionomer interferes with oxygen permeation to the catalytic surface.[2] Nevertheless, O_2 permeation can be increased either through higher diffusivity or higher solubility. However, these two parameters depend on porosity, tortuosity, and favorable O_2-Nafion interactions.[3] In this work we use molecular dynamics simulations to investigate oxygen permeability in bulk Nafion, and how these permeabilities are affected by polymer chain length and side chain moieties. Understanding these effects will help in the design of new perfluorosulfonic acid (PFSA) membrane materials.

References:

1. Ryosuke Jinnouchi, Kenji Kudo, Naoki Kitano, Yu Morimoto. *'Molecular Dynamics Simulations on O_2 Permeation through Nafion Ionomer on Platinum Surface'*. (2016) Electrochim. Acta 188: 767-776.
2. Yuya Kurihara, Takuya Mabuchi, Takashi Tokumasu. *'Molecular Simulation of Oxygen Permeation through Ionomer in Catalyst Layer'*. (2014) ECS Trans 64: 559-565.
3. Tomonori Kawakami, Isamu Shigemoto, Nobuyuki Matubayasi. *'Structure and permeability of ionomers studied by atomistic molecular simulation combined with the theory of solutions in the energy representation'*. (2018) J. Chem. Phys. 148, 214903.

596753 Topological Data Analysis: Applications to Soft Matter and Molecular Simulations

Abstract:

Data generated by experiments and complex simulations (e.g., molecular dynamics simulations) is often summarized using descriptive statistics (e.g., averages, moments, and correlation functions) in order to reduce complexity and to facilitate analysis [1]. Unfortunately, descriptive statistics might fail to capture important aspects of complex datasets. Specifically, statistical techniques might fail to capture key geometrical structures (e.g., complex heterogeneous domains). Interesting examples that illustrate these limitations are the anscombe quartet and the datasaurus dozen datasets [2,3]. These datasets are visually distinct (define different geometrical spaces) but have the exact same descriptive statistics (mean, standard deviation, and correlation).

Recent advances in applied topology and geometry have led to the development of a field known as Topological Data Analysis (TDA) [4,5,6]. TDA is a framework that views complex data through the lenses of geometry and topology. A particularly method of TDA, known as persistence homology, represents datasets (e.g., point clouds and images) as geometric objects and performs dimensionality reduction by projecting the data onto a low-dimensional space composed of elementary geometric objects (topological features) that persist at different scales [7,8]. The features are quantifiable and stable to basic deformations (e.g., stretching, rotation, bending) and can be used to perform different tasks (e.g., classification, regression) [9,10]. TDA methods have been applied successfully in materials science [11,12], time series and signal analysis [13,14], and bio-sciences [15,16].

This talk focuses on the application of TDA to complex datasets arising in soft matter and molecular dynamics (MD) simulations. We use TDA to characterize topological features that develop in liquid crystal films when exposed to air contaminants [17]. We also show how TDA can be used to characterize topological features of scatter fields for flow cytometry of emulsions containing liquid crystal droplets [18]. Finally, that TDA can be used to characterize the geometry of three-dimensional liquid-phase environments generated by MD [19]. For all of these studies, we show that the topological features are strongly correlated to emerging properties of interest (e.g., concentration of air contaminant or reactivity of a molecule in a solvent environment).

[1] Rahul R Shah and Nicholas L Abbott. Principles for measurement of chemical exposure based on recognition-driven anchoring transitions in liquid crystals. Science, 293(5533):1296–1299, 2001.

[2] Francis J Anscombe. Graphs in statistical analysis. The american statistician, 27(1):17–21, 1973.

[3] Justin Matejka and George Fitzmaurice. Same stats, different graphs: generating datasets with varied appearance and identical statistics through simulated annealing. In Proceedings of the 2017 CHI Conference on Human Factors in Computing Systems, pages 1290–1294, 2017.

[4] Gunnar Carlsson. Topology and data. Bulletin of the American Mathematical Society, 46(2):255–308, 2009

[5] Herbert Edelsbrunner and John Harer. Computational topology: an introduction. American Mathematical Soc., 2010.

[6] Afra Zomorodian. Topological data analysis. Advances in applied and computational topology,

70:1–39, 2012.

[7] Robert Ghrist. Barcodes: the persistent topology of data. Bulletin of the American Mathematical Society, 45(1):61–75, 2008.

[8] Gunnar Carlsson, Afra Zomorodian, Anne Collins, and Leonidas J Guibas. Persistence barcodes for shapes. International Journal of Shape Modeling, 11(02):149–187, 2005

[9] Fred´ eric Chazal, Vin De Silva, Marc Glisse, and Steve Oudot. The structure and stability of ′ persistence modules. arXiv preprint arXiv:1207.3674, 21, 2012.

[10] David Cohen-Steiner, Herbert Edelsbrunner, and John Harer. Stability of persistence diagrams. Discrete & Computational Geometry, 37(1):103–120, 2007

[11] Takashi Ichinomiya, Ippei Obayashi, and Yasuaki Hiraoka. Persistent homology analysis of craze formation. Physical Review E, 95(1):012504, 2017.

[12] Mickael Buchet, Yasuaki Hiraoka, and Ippei Obayashi. Persistent homology and materials in- ¨ formatics. In Nanoinformatics, pages 75–95. Springer, Singapore, 2018.

[13] Jose A Perea and John Harer. Sliding windows and persistence: An application of topological methods to signal analysis. Foundations of Computational Mathematics, 15(3):799–838, 2015.

[14] Bernadette J Stolz, Heather A Harrington, and Mason A Porter. Persistent homology of timedependent functional networks constructed from coupled time series. Chaos: An Interdisciplinary Journal of Nonlinear Science, 27(4):047410, 2017.

[15] Peter M Kasson, Afra Zomorodian, Sanghyun Park, Nina Singhal, Leonidas J Guibas, and Vijay S Pande. Persistent voids: a new structural metric for membrane fusion. Bioinformatics, 23(14):1753–1759, 2007.

[16] Hyekyoung Lee, Hyejin Kang, Moo K Chung, Bung-Nyun Kim, and Dong Soo Lee. Persistent brain network homology from the perspective of dendrogram. IEEE transactions on medical imaging, 31(12):2267–2277, 2012.

[17] Alexander Smith, Nicholas L Abbott and Victor M Zavala. Convolutional Network Analysis of Optical Micrographs. doi.org/10.26434/chemrxiv.11688924.v2.

[18] Shengli Jiang, JungHyun Noh, Alexander D Smith, Chulsoon Park, Nicholas L Abbott, Victor M Zavala. Identification of Endotoxins from Bacterial Species using Liquid Crystal Droplets and Machine Learning. Under Review, 2020.

[19] Theodore W Walker, Alex K Chew, Huixiang Li, Benginur Demir, Z Conrad Zhang, George W Huber, Reid C Van Lehn, and James A Dumesic. Universal kinetic solvent effects in acid-catalyzed reactions of biomass-derived oxygenates. Energy & Environmental Scient, 11(3):617-628, 2018.

610161 Evaluating the Adsorption Landscape for Polymers of Intrinsic Microporosity

Abstract:

The potential for microporous materials to provide energy-efficient solutions for chemical separations or storage has provided motivation for the development of many families of porous systems; for example, metal organic frameworks, covalent organic frameworks, zeolites, and microporous polymers. Many of these microporous materials are modular in nature or can be readily modified, which yields a substantial set of conceivable adsorbents. Studying these sets at variable operating conditions or for a diversity of adsorbates can be thought of as evaluating adsorption space. Molecular simulation screening studies provide a useful tool for this type of challenge because of their capability to efficiently predict adsorption properties for a large batch of systems, which can ultimately expedite adsorbent design/discovery. The presented study has applied a simulated screening approach for 15 unique polymers of intrinsic microporosity (PIMs) with 20+ distinct adsorbate species. To produce the full adsorption isotherms, a combination of molecular dynamics (MD) and Monte Carlo (MC) techniques are utilized, which greatly improves the accuracy of predictions made in the presence of appreciable adsorbate uptake by accounting for sorption-induced polymer chain rearrangement. This adsorption space screening study that incorporates polymer chain dynamics is the first of its kind and has become possible because of our newly developed python interface between RASPA and LAMMPS, which handle the Monte Carlo and Molecular Dynamics simulations, respectively. Additionally, to efficiently manage computational resources we have included a set of automated convergence statistics that allows the MC/MD process to terminate upon reaching the equilibrated state used for predictions, thus minimizing excess wall time. Using these methods, we have determined the onset swelling pressures for all of the adsorbate-adsorbent pairs, which is a quantity of industrial relevance. Finally, in an effort to circumvent laborious simulations, we examine the applicability of developing a model for the rapid prediction of adsorption properties in PIMs based on commonly calculated values, such as Henry's constant and helium void fraction.

609347 Simulation Workflows for Studying Adsorption and Transport at Patterned M1 Catalyst Surfaces

Abstract:

In this work we develop new computational tools to study the patterning of surfaces with applications in catalysis. The oxide Mo-V-Nb-Te-O (M1) can be an effective catalyst in the oxidative dehydrogenation of ethane and is the particular focus of the present work. While M1 holds promise to lower the cost of producing ethylene from ethane, it ages at the higher temperatures where conversion is highest. Surface patterning strategies aim to facilitate transport of reagents to active sites on the catalyst surface while stabilizing the catalyst itself. Here we use molecular simulations to investigate how the patterning of the surface of M1 with organic molecules influences adsorption isotherms and surface transport of ethane and ethylene. We use tooling developed within the Molecular Simulation Design Framework (MoSDeF) to solve common molecular simulation tasks, including: Initializing independent initial configurations of M1 surfaces patterned with different molecules, combining force fields, and handing off configurations between Monte Carlo and molecular dynamics engines to perform equilibrium and transport calculations, respectively. We develop analysis tools for measuring adsorption near the reactive Vanadium and Tellurium sites and discuss strategies for increasing ethane adsorption preference towards Vanadium. Using this tooling we automate the performance of simulation workflows comprising a set of planar organic surface modifiers and temperatures. We identify surface modifiers and temperatures that are expected to improve catalyst selectivity and yield and discuss challenges and opportunities towards experimental implementations.

609305 Combining Neural Network Adsorption Predictions with Free Energy Calculations: A New Framework for Computational MOF Discovery for Chemical Separations

Abstract:

High throughput computational screening (HTCS) of metal-organic frameworks (MOFs) for adsorption applications using molecular simulation has become standard over the last decade. Notable examples, where screening has led to synthesis of a high-performing MOF, include screenings for methane and hydrogen storage. However, widespread efficacy of traditional HTCS is limited for two reasons: first, while usually faster than experiments, molecular simulation can become prohibitively expensive as the number of MOFs and operating conditions increase. Second, experimental synthesis of MOFs from computational databases is rare due to uncertainty as to whether computationally built MOF prototypes can be experimentally synthesized. Here we begin by illustrating how simulated data of the adsorption of alchemical species can be used to train neural networks to almost instantly predict full adsorption isotherms for diverse real adsorbates in MOFs. Specifically, we present an extension to our existing neural network trained to predict single-component adsorption isotherms [1] that allows for the rapid prediction of binary adsorption isotherms for small-gas adsorbates in diverse MOFs. Second, we show how MOF free energy can be calculated on a large scale using thermodynamic integration and how these free energies can be used to eliminate synthetically unlikely MOFs from a computational database [2]. Finally, as a proof-of-concept for this new screening framework, we use binary adsorption data generated using our extended neural network model to isolate the most promising MOFs for the separation of Kr/Xe mixtures from a topologically diverse MOF database. Then, we eliminate synthetically unlikely MOFs based on their relative free energies. We demonstrate that by using this combined approach, not only can we screen a much larger number of MOFs and operating conditions than was previously possible, but we can also gain insight into which among the most promising MOFs are also promising synthetic targets.

[1] Anderson, R.; Biong, A.; Gómez-Gualdrón, D. A. Adsorption Isotherm Predictions for Multiple Molecules in MOFs Using the Same Deep Learning Model. *J. Chem. Theory Comput.* **2020**.

[2] Anderson, R.; Gómez-Gualdrón, D. A. Large-Scale Free Energy Calculations on a Computational MOF Database : Toward Synthetic Likelihood Predictions. *ChemRxiv* **2020**, 1–16.

597575 Machine Learning Using the Guest/Host Energy Histogram to Predict the Adsorption of Chain Molecules

Abstract:

Machine learning, as it gains more and more usefulness in computational research, has been successfully used to study nanoporous materials such as zeolites and metal-organic frameworks (MOFs). Recently, Bucior et al. published a study[1] using histograms of the guest/host energy as features for LASSO, which is a linear regression machine learning method, to investigate and predict the top candidates for hydrogen and methane storage in MOFs. In this work, we refined the energy histogram methodology and extended it to longer alkanes such as ethane, propane, and n-hexane. Energy histograms are now automatically aggregated without pre-determined bin sizes such that if a range of energy values has more counts, it is represented with more bins in the histogram. To test our new methodology, we performed grand canonical Monte Carlo (GCMC) simulations for materials from the ToBaCCo set of MOF structures[2]. Both linear and non-linear models such as LASSO,[3] Random Forest,[4] and Neural Nets (NN)[5] were used to develop machine learning models to predict adsorption. We find that the size of the molecule plays a big role in determining which ML methods work best. For ethane, simple linear regression (LASSO) works well, but for propane and hexane we must turn to non-linear methods (RF and NN).

For n-hexane, we found a large number of outliers in the parity plot of ML versus GCMC. We determined that these outliers correspond to points on the adsorption isotherm near the point of condensation in the pores. Through machine learning, we were able to find that the GCMC simulations had not converged. Thus, the error was in the simulation and not the ML model. These un-converged points are nearly impossible to identify by looking at error bars generated from GCMC simulations because the systems remain trapped in one state and appear converged. Thus, we discovered a way to improve the quality of high-throughput screening using molecular simulations at these critical conditions through machine learning.

We also tested additional new features to compensate information loss when constructing one-dimensional histograms from the three-dimensional energy grids. Additionally, the new energy histogram algorithm shows robustness when predicting selectivity for Xe/Kr separation.

References:

(1) Bucior, B. J.; Bobbitt, N. S.; Islamoglu, T.; Goswami, S.; Gopalan, A.; Yildirim, T.; Farha, O. K.; Bagheri, N.; Snurr, R. Q. Energy-Based Descriptors to Rapidly Predict Hydrogen Storage in Metal–Organic Frameworks. *Mol. Syst. Des. Eng.* **2019**, *4* (1), 162–174. https://doi.org/10.1039/C8ME00050F.

(2) Colón, Y. J.; Gómez-Gualdrón, D. A.; Snurr, R. Q. Topologically Guided, Automated Construction of Metal–Organic Frameworks and Their Evaluation for Energy-Related Applications. *Cryst. Growth Des.* **2017**, *17* (11), 5801–5810. https://doi.org/10.1021/acs.cgd.7b00848.

(3) Tibshirani, R. Regression Shrinkage and Selection via the Lasso. *J. R. Stat. Soc. Ser. B Methodol.* **1996**, *58* (1), 267–288.

(4) Liaw, A.; Wiener, M. Classification and Regression by RandomForest. **2002**, *2*, 5.

(5) Abadi, M.; Barham, P.; Chen, J.; Chen, Z.; Davis, A.; Dean, J.; Devin, M.; Ghemawat, S.; Irving, G.; Isard, M.; Kudlur, M.; Levenberg, J.; Monga, R.; Moore, S.; Murray, D. G.; Steiner, B.; Tucker, P.; Vasudevan, V.; Warden, P.; Wicke, M.; Yu, Y.; Zheng, X. TensorFlow: A System for Large-Scale Machine Learning. In *12th USENIX Symposium on Operating Systems Design and Implementation (OSDI 16)*; 2016; pp 265–283.

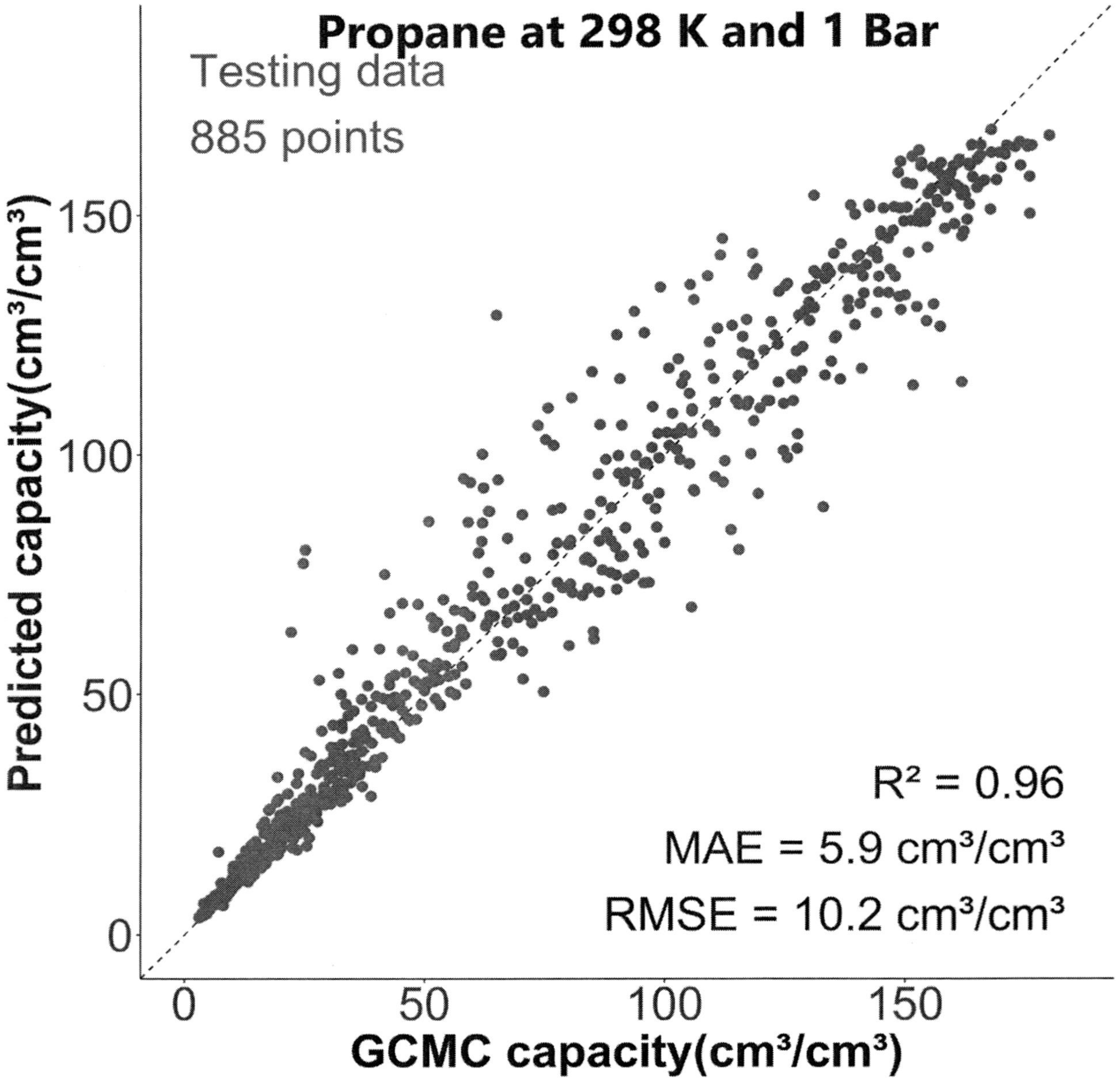

607377 Comparing the Performances of an Ionic Liquid-Based Absorption Cycle and an Organic Rankine Cycle for Power Generation from Low-Grade Waste Heat

Abstract:

The Organic Rankine Cycle (ORC) has been employed over the past half-century to generate power using lower temperature (lower-grade) heat sources than the typical steam temperatures in the common water-based Rankine cycle. While often advertised as requiring only "waste" heat, most commercial ORC units require heat sources well in excess of 100 °C, while extensive amounts of waste heat near 100°C are often vented to the atmosphere by industry. Absorption power cycles (APCs) have been proposed to overcome these deficiencies of the ORC, especially when using low-grade waste heat sources at temperatures of 100 °C or lower. In this work, the performance of an APC using low-grade waste heat and a mixture of the ionic liquid (IL) 1-hexyl-3-methylimidazolium bis(trifluoromethylsulfonyl)imide ([HMIm][Tf2N]) with the common fluorinated refrigerant R-134a was compared against an ORC using R-134a and operating at similar conditions. Both cycles were modeled in Aspen Plus using two thermodynamic models, the Peng-Robinson equation of state with parameters fitted to vapor-liquid equilibrium data of our working fluid mixtures, and the COSMO-SAC purely predictive model. Our results show that the IL-based APC has higher efficiencies than the ORC, with the gap increasing as the temperature of the waste heat sources is lowered. The effects of the cycle operational temperatures on the efficiency and flow rates of working fluid will be presented and discussed.

610591 GPU Optimized Monte Carlo Version 2.50

Abstract:

GOMC is a general-purpose Monte Carlo simulation engine for the simulation of molecular systems with molecular mechanics force fields based on the 12-6 Lennard-Jones, or Mie potentials[1]. It has support for simulations in all common ensembles, including the Gibbs ensemble Monte Carlo algorithm. GOMC was designed with a focus on high performance, and has support for simulations on multi-core CPUs and graphics processing units (GPUs). This poster highlights a number of recent enhancements to GOMC, including new Monte Carlo moves to enhance the sampling of phase space, such as Molecular Exchange Monte Carlo (MEMC)[2,3], configurational-bias for molecules that contain rings, the crankshaft move, parallel tempering, force/torque-biased multi-particle move as well as a multi-particle move using Brownian dynamics[4]. Support for force fields governed by exp-6 potentials, and free energy calculations using thermodynamic integration or free energy perturbation has been added.

[1] Nejahi, Y, Barhaghi, MS, Mick, J, Jackman, B, Rushaidat, K, Li, YZ, et al. *GOMC: GPU Optimized Monte Carlo for the simulation of phase equilibria and physical properties of complex fluids.* Softwarex, 2019; **9**: 20-7.

[2] Barhaghi, MS, Torabi, K, Nejahi, Y, Schwiebert, L,Potoff, JJ. *Molecular exchange Monte Carlo: A generalized method for identity exchanges in grand canonical Monte Carlo simulations.* J. Chem. Phys., 2018; **149**(7): 072318.

[3] Barhaghi, MS,Potoff, JJ. *Prediction of phase equilibria and Gibbs free energies of transfer using molecular exchange Monte Carlo in the Gibbs ensemble.* Fluid Phase Equilib., 2019; **486**: 106-18.

[4] Moucka, F, Rouha, M,Nezbeda, I. *Efficient multiparticle sampling in Monte Carlo simulations on fluids: Application to polarizable models.* J. Chem. Phys., 2007; **126**(22): 224106.

611248 Open Force Field Initiative: New Strategies for Parameterizing Non-Bonded Interactions

Abstract:

The Open Force Field Initiative (OpenFF) is a collaborative effort from academia and industry focused on delivering better biomolecular force fields through innovative science and data curation. One of the major areas we aim to improve is the treatment of van der Waals interactions. These interactions are crucial to modeling many biologically relevant processes, like protein folding, protein-ligand binding, and membrane transport. Modeling of these interactions is difficult; they are generally fitted against condensed-phase property data, which creates a "black box" problem where the relationship between model parameters and fitting targets is not entirely clear.

The development of the OpenFF Evaluator software package, coupled with datasets sourced from the NIST ThermoML database, has accelerated our fitting process dramatically. This software provides a framework for streamlined property estimation, fitting and benchmarking. Heat of vaporization is a common fitting target for vdW models, but it is problematic, due to differences in polarization between the liquid and gas phase, as well as limited data availability. Using the Evaluator software, we fit vdW interactions against combinations of pure density, heat of vaporization, mixture density, and enthalpy of mixing for a variety of organic species. Results show that fitting against mixture properties can yield improvements in vdW modeling, and that enthalpy of mixing data may be an effective replacement for heat of vaporization.

We also explore methods to rationally choose models for vdW interactions. In addition to continuous choices about models, there are also discrete choices of functional forms: number of atom types, combination rules, and vdW functional forms. We demonstrate the use of Bayesian inference as a tool to rationally select between different models, or levels of complexity. For a simple 2CLJQ fluid model, Bayesian inference gives evidence that, for selected physical properties, a quadrupole interaction is unnecessary. We then discuss strategies and progress towards building surrogate models that will enable this analysis for more complex biomolecular force fields.

604142 How Can Machine Learning Accelerate the Sampling and Interpretation of Molecular Dynamics Simulations?

Abstract:

Molecular dynamics simulations offer high spatial and temporal resolutions to decipher functional mechanisms of membrane proteins. Major improvements in hardware, algorithms, and computational resources have made it possible now to investigate larger and more realistic protein systems. However, the capabilities of molecular dynamics simulations are still severely limited by three major problems. First, only a few crystal structures are available for membrane proteins and these structures are mostly restricted to key metastable states that the protein adopts. Second, membrane protein simulations require access to large computational resources, of the order of several 100,000 hours on a single Nvidia GTX1080 GPU using a popular molecular simulation package such as NAMD. Finally, it is difficult to extract valuable insights from the resulting high-dimensional simulation data (several terabytes).

In this study, we aim at developing efficient computational tools to accelerate the sampling and interpretation of molecular dynamics simulations. To address the absence of structural information, we developed a machine learning based algorithm (FingerprintContacts) to quickly predict multiple protein structures by combining agglomerative clustering and co-evolutionary information. We have demonstrated the capabilities of FingerprintContacts on eight proteins with varying conformational motions. To enhance the sampling efficiency, we proposed that evolutionary couplings can be used as reaction coordinates to efficiently guide the sampling of complex conformational free energy landscapes. To interpret the resulting high-dimensional simulation data, we developed a genetic algorithm based method to automatically select features for dimensionality reduction. The integration of the developed algorithms and all-atom molecular dynamics simulations has allowed us to characterize long timescale conformational transitions and the complete substrate translocation cycle of two nitrogen transporters. This work would establish efficient computational frameworks for understanding long timescale biophysical processes.

602207 Utilizing Parallel Biasing Methods to Understand Host-Guest Systems

Abstract:

Host-guest systems, where a small guest molecule resides within a pocket of a larger host molecule, are of interest in many applications, including drug delivery, toxin removal, membrane formation, hydrogels, biosensing, and many developing areas of materials science. Due to the broad variety of potential host-guest systems for a given application, molecular engineering through the computational study of host-guest binding is an attractive possibility. In this study, we explore how new "path-agnostic" algorithms, such as parallel biasing metadynamics compare with more standard "path-based" methods such as thermodynamic integration in the study of host-guest binding. This is an important question, as parallel biasing methods have particular promise for the study of binding in complicated host-guest systems, particularly those with multiple binding modes, due to their ability to efficiently explore multiple reaction coordinates simultaneously. In studying the binding affinity and other binding characteristics in systems such as cucurbit-7-uril with fentanyl, we compare our results with experiment and discuss the importance of collective variable choice in binding calculations.

607298 Virial Coefficients As a Route from Molecular Models to Accurate Thermodynamic Models

Abstract:

Virial coefficients can provide a thermodynamic model that is exactly correct for a given molecular model. It is limited in state space (non-condensed, low density), but its accuracy can be self-assessed by examining convergence. Recent developments in the quality of molecular models, both semi-empirical and ab initio, provide a foundation to formulate thermodynamic models this way. Additionally, improvements in methodology allow for calculation of virial coefficients and their derivatives, particularly for non-pairwise molecular models (which are needed for highly accurate modeling).

We demonstrate with two types of applications: (a) Using a semi-empirical model, we generate an equation of state for a multicomponent mixtures, with exact description of composition dependence. This won't yield results better than experiment, but it is preferable to other thermodynamic models at conditions where it is converged. We pursue a systematic, comprehensive effort to apply these methods to TraPPE potentials for a collection of 4 species - nitrogen, oxygen, carbon dioxide and ammonia and their mixtures. (b) Using highly accurate ab initio potentials, we generate an equation of state that exceeds the accuracy of what is possible experimentally. We examine the accuracy of ab intio virial coefficients for He-4 which are based on accurate first-principles 2- and 3-body molecular models from the literature and are generated using path-integral Monte Carlo (PIMC) calculations. We show that these coefficients can be used make predictions of pressure which are consistent with, and more precise than those obtained from experiment. We also compare the predictions of speed of sound against those from experiment.

611153 Bayesian Model Selection for Non-Covalent Interactions

Abstract:

Non-covalent interactions, commonly divided into electrostatic and van der Waals interactions, play critical roles in many molecular processes. It is crucial to model these interactions accurately when performing molecular dynamics (MD) simulations. Non-covalent interactions are generally difficult to parameterize, because relatively simple models must describe complex interactions between many particles. We focus on the modeling of van der Waals interactions, which require decisions about functional forms, parameters, combining rules and atom typing.

As part of the Open Force Field Initiative, we are developing force fields, including non-covalent interaction models, using data-driven techniques. To this end, we explore the use of Bayesian inference to make data-driven choices between dispersion-repulsion parameters and functional forms, by calculating *Bayes factors*, which are essentially 'odds' between different models and sets of parameters. This strategy requires repeated evaluation of parameter sets, which has previously been a large source of computational expense.

In this study, we test this strategy on the 2-center Lennard Jones plus Quadrupole (2CLJQ) model for simple fluids, as its simple functional form is easily modified and analytical 'surrogate models' exist in the literature, allowing for fast evaluation of parameter sets. In this way, we can sample over the entire distribution of parameters and calculate Bayes factors, without incurring the computational cost of equilibrium simulations. Using the reversible jump Monte Carlo (RJMC) algorithm, we sample the posterior probability distributions of both the *models* and the *parameters.*

We ask whether including the model's quadrupole parameter is justified while reproducing temperature-dependent density, saturation pressure, and surface tension data for simple molecules. In general, we find that the quadrupole is not justified for reproducing these properties (with several notable exceptions). Additionally, we produce parameter probability distributions for these compounds, valuable information for guiding future parameterization; through these distributions we identify targets for future dimensional reduction. This work demonstrates the utility of Bayesian inference as a tool for model selection and paves the wave for future application of this technique to more complex decisions required in fitting complete force fields.

607000 Parallel Prefetching in Canonical and Grand Canonical Ensemble Monte Carlo Simulations

Abstract:

In order to enable large-scale molecular simulations, algorithms must efficiently utilize multi-core processors that continue to increase in total core count, but with relatively stagnant clock speeds. Although parallelized molecular dynamics (MD) software has taken advantage of this trend in computer hardware, single-particle perturbations with Monte Carlo are more difficult to parallelize than system-wide updates in MD using domain decomposition. Instead, prefetching reconstructs the serial Markov chain after computing multiple Monte Carlo trials in parallel. Canonical ensemble simulations of a Lennard-Jones fluid with prefetching resulted in up to a factor of 1.7 speedup using 2 OMP threads, and a factor of 3 speedup using 4 threads. Flat histogram grand canonical ensemble simulation see further improvement with up a factor of 3.5 speedup using 4 threads. Strategies for maximizing efficiency of prefetching simulations are discussed, including the potentially counter-intuitive benefit of reduced acceptance probabilities. Determination of the optimal acceptance probability for a parallel simulation is simplified by theoretical prediction from serial simulation data. Source code is available in the prefetch plugin of the Free Energy and Advance Sampling Simulation Toolkit (FEASST).

609760 Development of a Novel Method for Upscaling Molecular Dynamics to Coarse-Grained Models

Abstract:

Recent scientific work has shown that subsurface deposits of hydrocarbons in shale exhibit thermodynamic and flow behavior inconsistent with equations of state for bulk mixtures. Understanding fluid properties ranging from the nanometer to micron scales is important in enhanced oil recovery as well as other mesoscale flow systems. Molecular Dynamics (MD) simulation is an indispensable tool for determining the characteristics of confined fluid systems such as those encountered in shale oil deposits. However, performing MD computations for micron scale systems is currently not feasible on most computational resources. One means of overcoming this challenge is the use of Dissipative Particle Dynamics (DPD), a semi-empirical class of models specifically targeting mesoscale systems. Proper application of DPD models requires calibration of a number of parameters from data obtained experimentally or through MD simulations.

In this work, we establish a method of calibrating DPD model parameters from MD data, specifically, by computing a set of local thermodynamic and transport properties and optimizing DPD parameters from spatially filtered properties. MD datasets for both bulk heptane and heptane confined in a silica pore are considered. Additionally, we assess several functional forms for the DPD force kernel by comparing their *a priori* and *a posteriori* predictions of system properties.

608695 CG Pyrosetta: A Coarse-Grained Configurational Search Tool for General Heteropolymers

Abstract:

Foldamers are a newly emerging class of macromolecules made up from a set of diverse non-natural monomer subunits. Similar to proteins, foldamer macromolecules show promise for highly specific, designable materials. Currently, there is a limited collection of structural data for these foldamer macromolecules, making traditional protein-based folding techniques unavailable to these types of molecules. In this work, we propose a hypothesis-based coarse-graining (CG) approach to explore the available secondary structure of general foldamer molecules. To this end, we developed CG PyRosetta, an extension to Rosetta, the widely used protein folding software, to explore the secondary structure available to arbitrary foldamer macromolecules. CG PyRosetta can fold arbitrary linear backbones, and can, *ab initio,* identify secondary structure elements in simple CG models provided a parameter set. Attaching hypotheses to these CG parameters, we develop a set of folding trends observed from the emergent secondary structures present in these models. By applying these folding principles to experimental monomers, we can guide experimental design of new foldamers. We can use this framework to explore the effects of sidechain size, monomer rigidity and sequence heterogeneity on the formation of helices and other secondary structures.

606412 Combining Strategic Training Data Selection and Feature Engineering to Reach Accurate and Efficient Molecular Property Prediction

Abstract:

Organic molecular design problems, such as drug discovery or material design, aim to identify molecules with desired properties from the chemical space, wherein the number of potential compounds is estimated to reach 10^{60}. The size of the chemical space forbids experiments or high-level quantum chemistry to evaluate each molecule. In recent decades, the integration of machine learning methods with virtual screening makes the exploration of chemical space practical due to its high efficiency and low cost. While many machine learning models manage to reach high accuracy with hundreds of thousands of training molecules, only a handful of study has been focused on optimizing the model performance under a tight computation budget. In this work, we propose a strategy to obtain accurate machine learning predictions with a minimum number of data points required for training. Specifically, we address the problem in threefold. First, we demonstrate the efficacy of a method that adaptively builds the compact training set by systematically balancing exploitation via experimental design and exploration of the space via cheminformatics-based diversity maximization procedures. Second, we expand this procedure with the use of nonlinear and locally linear dimensionality reduction methods to leverage data embeddings. Third, we focus on improving the model accuracy under the constraint of a small training set, which we achieve by progressively incorporating nonlinearity to our modified group additivity approach.

608731 Gmso: A Flexible, Python-Based Representation of Chemical Topologies for Molecular Simulation

Abstract:

The Molecular Simulation Design Framework (MoSDeF)[1] is a Python-based open-source software suite focused on constructing, atom typing, and parametrizing chemical systems. The package is architectured to provide researchers the ability to prepare chemical systems in a scriptable manner, crucial to enhancing the reproducibility of computational science. The suite comprises three main libraries: mBuild[2,3] which handles the construction of chemical systems, foyer[4,5] which is responsible for automatically atom typing and parametrizing chemical systems, and the General Molecular Simulation Object (GMSO)[6], which serves as the backend data structure that encapsulates system and forcefield information, which can be written out to various file formats specific to simulation engine. GMSO is designed with the purpose of providing a more dynamic and flexible molecular topology data structure, compared to other open-source libraries currently available, including those previously used in MoSDeF. In order to make the library as general as possible, GMSO makes no assumptions regarding the details of molecular systems, including the functional forms of the potential energy equations, and the unit system. The unyt[7,8] library is implemented to manage the internal unit system, while interatomic potential expressions are handled by the sympy[9,10] library, and both can be explicitly defined by the users. This enables users to either choose from pre-built templates GMSO provides or create more exotic non-standard functional forms. While the use of non-standard potentials may be limited by the compatibility with current simulation engines, the benefit of having a flexible backend will allow MoSDeF to quickly support future potential forms. Many engines support these more exotic potential forms through the use of tabulated potentials, which are also supported by GMSO. Furthermore, GMSO defines a new, more explicit XML schema for force fields, further improving on the force field schema implemented previously in foyer. The new form of XML provides the same information as the original forcefield format defined in foyer, including parameters and usage rules, but further includes information regarding units and the explicit functional form of the given interaction potential. By doing so, the use of GMSO and MoSDeF reduces ambiguity of the force field form, parameters, and usage, and further supports the concept of Transparent, Reproducible, Usable By Others, and Extensible (TRUE)[11] simulations.

References

1. "MoSDeF" [Online]. Available: https://github.com/mosdef-hub.
2. C. Klein, J. Sallai, T. J. Jones, C. R. Iacovella, C. McCabe, and P. T. Cummings, "A Hierarchical, Component Based Approach to Screening Properties of Soft Matter", Foundations of Molecular Modeling and Simulation, **2016**, pp. 79-92.
3. "mBuild" [Online]. Available: https://github.com/mosdef-hub/mbuild.
4. Christoph Klein, Andrew Z. Summers, Matthew W. Thompson, Justin B. Gilmer, Clare McCabe, Peter T. Cummings, Janos Sallai, Christopher R. Iacovella. "Formalizing atom-typing and the dissemination of force fields with foyer", Computational Materials Science, Volume 167, **2019**. https://doi.org/10.1016/j.commatsci.2019.05.026.
5. "foyer" [Online]. Available: https://github.com/mosdef-hub/foyer.
6. "gmso" [Online]. Available: https://github.com/mosdef-hub/gmso.
7. Goldbaum et al., (2018). unyt: Handle, manipulate, and convert data with units in Python. Journal of Open Source Software, 3(28), 809, https://doi.org/10.21105/joss.00809

8. “unyt” [Online]. Available: https://github.com/yt-project/unyt.
9. Meurer A, Smith CP, Paprocki M, Čertík O, Kirpichev SB, Rocklin M, Kumar A, Ivanov S, Moore JK, Singh S, Rathnayake T, Vig S, Granger BE, Muller RP, Bonazzi F, Gupta H, Vats S, Johansson F, Pedregosa F, Curry MJ, Terrel AR, Roučka Š, Saboo A, Fernando I, Kulal S, Cimrman R, Scopatz A. 2017. SymPy: symbolic computing in Python. *PeerJ Computer Science* 3:e103. https://doi.org/10.7717/peerj-cs.103
10. “sympy” [Online]. Available: https://github.com/sympy/sympy.
11. Matthew W. Thompson, Justin B. Gilmer, Ray A. Matsumoto, Co D. Quach, Parashara Shamaprasad, Alexander H. Yang, Christopher R. Iacovella, Clare McCabe & Peter T. Cummings (2020) Towards molecular simulations that are transparent, reproducible, usable by others, and extensible (TRUE), Molecular Physics, DOI: 10.1080/00268976.2020.1742938

608909 Exact Sampling of Stochastic Process from Brownian Bridge

Abstract:

Simulation of macromolecular processes towards the study of phase transitions is seriously deterred by extreme computational tortuosity. We report here a methodology based on the concept of a Brownian bridge that has the promise to contribute significantly to alleviate this problem. We discuss our approach with examples from the statistics of a continuous polymer chain in an external field, a canonical problem of the molecular processes that has tremendous applications in materials science and biology, and is fundamental in field theories describing the phase behavior of concentrated melts and solutions. Using a Brownian bridge (a stochastic process with its start point and end point specified), one can systematically (i) design polymer chains of a given topology, (ii) sample polymer chains in a complicated energy landscape, i.e., a landscape with metastable states separated by large free-energy barriers, and (iii) sample rare conformational states in phase space. Furthermore, one can extend this formalism to condition a polymer to lie a particular region of phase space for its entire path, not just the end points. We note that the bridge formalism is not hampered by restrictive assumptions that are characteristic of other approaches such as Monte Carlo methods and importance sampling techniques such as umbrella sampling. Further, it navigates the process along specific paths by continually employing corrective drifts to the stochastic trajectories. Thus, future applications may be envisaged for computing the so-called minimum free energy paths commissioning the use of the string method in collective variables to study phase transitions. Finally, we propose to solve the Brownian bridge in a high dimensional space for the modeling of complex polymer chains (e.g., semi-flexible polymers).

598352 Data and Information Management for Pharma 4.0

Abstract:

As industry marches towards Pharma 4.0, data and information management becomes central to a robust data strategy. This strategy includes data capture, automation, analytics, modeling and knowledge management, which all rely on connected, free-flowing information. Given the wide variety of data types and sources, managing these flows is a great challenge. By leveraging modern technologies, digitization and culture enhancement, these complex flows can be streamlined into robust data pipelines that provide faster, wider access to data and knowledge that supports quicker and more informed decisions.

Data pipelines and integrations have many practical uses. One such use is that when data is readily available, models can have access to the most recent and relevant inputs, like new or updated material properties. Another use is the automated capture and recording of lab testing. An example pipeline will be shown that highlights the connection of lab testing, a material property database and updated model results.

Areas of interest include structured and unstructured data capture, automation, dashboarding, democratization of models, and knowledge tracking.

Author is an employee of AbbVie and may own AbbVie stock. AbbVie sponsored and funded the study; contributed to the design; participated in the collection, analysis, and interpretation of data, and in writing, reviewing, and approval of the final publication.

600128 Differential Gene Set Enrichment Analysis: A Statistical Approach to Quantify the Relative Enrichment of Two Gene Sets

Abstract:

Gene Set Enrichment Analysis (GSEA) is an algorithm widely used to identify enrichments of individual gene sets in transcriptomic data. However, statistical methods are needed to accurately measure how two gene sets or pathways are coordinately regulated with respect to each other. Here we present Differential Gene Set Enrichment Analysis (DGSEA), an adaptation of GSEA that assesses the relative enrichment of two gene sets. Using the metabolic pathways glycolysis and oxidative phosphorylation as an example, we demonstrate that DGSEA accurately captures the hypoxia-induced shift towards glycolysis. We also show that DGSEA is more predictive than GSEA of the metabolic state of cancer cell lines, including lactate secretion and intracellular concentrations of lactate and AMP. Furthermore, we demonstrate that DGSEA identifies novel metabolic dependencies not found by GSEA in cancer cell lines. Together, these data demonstrate that DGSEA is a novel tool to examine the relative enrichment of two gene sets. DGSEA is available as an R package at https://github.com/JamesJoly/DGSEA and can be readily applied to analyze gene expression data.

606278 Understanding the Chemical Machine Learning Design Space Using a Property Graph Database

Abstract:

Machine learning has been used extensively to predict molecular properties and design molecules [1-6]. When initializing a machine learning model, the modeler must make a series of decisions about how the model will operate. One must choose a learning algorithm (i.e random forest vs neural network), a featurization method (how the molecules in the data set will be described to the learning algorithm), training set size, hyper-parameters values, validation method, etc. These decisions are not independent and impact the cost and efficacy of the machine learning model. In this work, we trained a series of machine learning models using a wide gamut of the above parameters. For example, one instance could be a random forest model with 10-fold cross validation and Morgan molecular fingerprint featurization to predict logP. Each model was used to predict molecular properties and was evaluated based on error and prediction uncertainty. The model parameters, performance, and molecular datasets were stored in a property graph database (PGDB). Graph topology algorithms were used to identify model features, including molecular fragments, that most impact a model's performance. The PGDB enhances the explainability of machine learning models by enabling visualization and efficient queries of relationships between modeling choices, data, and model performance.

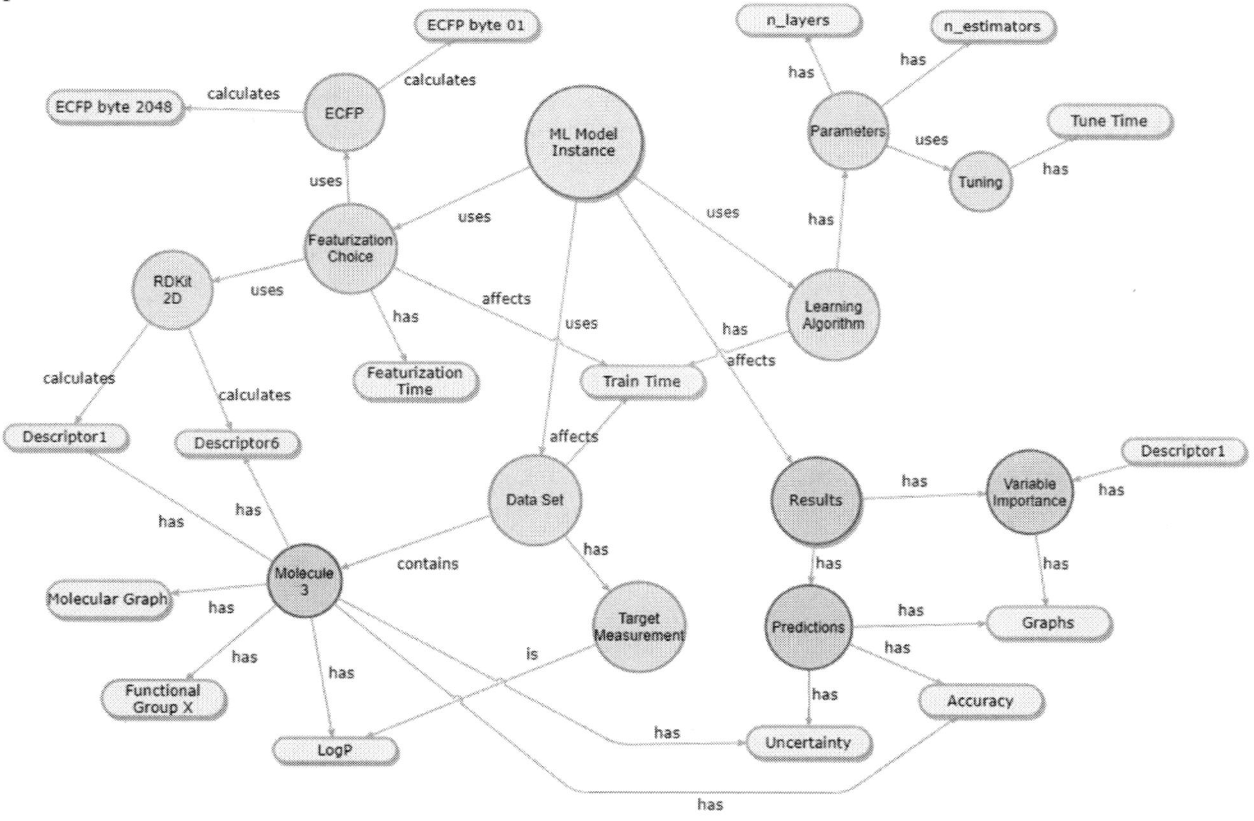

601957 Prediction of Solvation Free Energies Using PC-SAFT Classical Density Functional Theory

Abstract:

The calculation of the solvation free energy (SFE) of complex solutes is an important task in technical and biological systems. The calculations are routinely performed with molecular dynamics (MD) or Monte Carlo simulations and the SFE is evaluated with off-the-shelf methods like BAR, MBAR and umbrella sampling. However, the estimation of the SFE is computationally expensive requiring large enough systems for suffient sampling of multiple points along the thermodynamic integration path. Classical density functional theory (DFT) reduces the computational cost but contains the full molecular fluid structure at the solute-fluid interface. Thus enabling the rapid sampling of large numbers of possible solute/solvent combinations. The previously published molecular density functional theory [1] is based on a six-dimensional solvent density and its free-energy functional is based on the homogeneous reference fluid approximation but requires the solvent‘s bulk direct correlation function as an input.

The PC-SAFT equation of state has succesfully been applied to inhomogeneous systems at solid-fluid interfaces [2] and the frameworks is expanded to study solvation in non-polar solvents. The predictive nature of the PC-SAFT equation of state allows the calculation of SFEs without any adjustable parameters or the prior knowledge of the solvent‘s direct correlation function.

Our works shows excellent agreement between the SFEs calculated by PC-SAFT DFT and MD simulations for a wide range of solute/solvent combinations in broad temperature and pressure windows. Furthermore, we calculate self-SFEs, i.e. residual chemical potentials, in different solvent mixtures and obtain very good results compared to the PC-SAFT equation of state. A current field of interest is the extension to polar solvents like alcohols and water.

[1]: Jeanmairet, G., Levesque, M., Vuilleumier, R., & Borgis, D. (2013). Molecular density functional theory of water. *The journal of physical chemistry letters*, *4*(4), 619-624.

[2]: Sauer, E., Terzis, A., Theiss, M., Weigand, B., & Gross, J. (2018). Prediction of contact angles and density profiles of sessile droplets using classical density functional theory based on the PCP-SAFT equation of state. *Langmuir*, *34*(42), 12519-12531.

609051 Prediction of Structure and Charge Transport in Itic-Derived Electron Acceptors for Plastic Solar Cells

Abstract:

The goal of this project is to help experimentalists choose molecules and conditions optimized for synthesizing high-efficiency solar cells made from organic components. The structure and packing of molecules in organic photovoltaics (OPVs) influences charge transport and overall solar cell efficiency. We develop new software for initializing, simulating, and analyzing the morphologies of candidate OPV compounds. Here we focus on using molecular dynamics (MD) simulations to predict the thermodynamically stable morphologies of electron acceptor molecules ITIC, ITIC-F4, and CZTPTZ8FITIC. The equilibrated molecular packings from the MD simulations are used to predict charge transport properties and for model validation when experimental measurements are available. We discuss the thermodynamic conditions that we observe correlate with assembly of morphologies with higher charge transport. We find that these three ITIC-derived electron acceptors assemble structures with low long-range order and that their favorable charge transport properties may arise from the isotropic network of pi-stacked connections between molecules.

606369 Molecular Simulation Studies of Solvent-in-Salt Electrolytes for Energy Storage Applications

Abstract:

Significant efforts have been made to improve the performance of capacitive energy storage devices, which exhibit high power density but low energy density. Traditional organic electrolytes display relatively high conductivities, but low electrochemical stabilities which negatively impact the rate handling capabilities of these energy storage devices. Ionic liquids have been extensively studied in recent years due to their high electrochemical stability, but suffer from slow dynamical properties. Solvent-in-salt electrolytes (SISE), defined as highly concentrated salt solutions, have recently emerged as promising electrolytes for capacitive energy storage devices due to having high electrochemical stabilities while also displaying relatively high conductivity[1].

Here we present an integrated study of SISE[2] through various computational and experimental approaches to better understand the molecular behavior that drives dynamical properties. Specifically, neutron scattering and NMR experiments are performed to measure ion diffusivity, while impedance spectroscopy is conducted to measure the ionic conductivity of these electrolytes. Accompanying these experiments is a series of Molecular Dynamics (MD) simulations which provide an analysis of ion pairing and structuring to better understand the atomistic origins of ionic conductivity. Furthermore, the role of confinement on the dynamical properties is investigated through studies of SISE in novel MXenes[3], which are a family of two-dimensional metal carbides and nitride materials. Included are Grand Canonical Monte Carlo (GCMC) and Gibbs Ensemble Monte Carlo (GEMC) simulations to better understand how SISE fill and interact with the interlayer spacings of MXenes.

1. Suo, L., Hu, Y.S., Li, H., Armand, M., Chen, L. (2013). A new class of Solvent-in-Salt electrolyte for high-energy rechargeable metallic lithium batteries. *Nature Communications*. 4(1481).
2. Popov, I., Sacci, R. L, Sanders, N. C., Matsumoto, R. A., Thompson, M. W., Osti N. C., Kobayashi, T., Tyagi, M. S., Mamontov, E., Pruski, M., Cummings, P. T., Sokolov, A. (2020). Critical Role of Anion-Solvent Interactions for Dynamics of Solvent-in-Salt Solutions. *Journal of Physical Chemistry C*. https://doi.org/10.1021/acs.jpcc.9b10807
3. Anasori, B., Lukatskaya, M. R., Gogotsi, Y. (2017). 2D metal carbides and nitrides (MXenes) for energy storage. *Nature Review Materials*. 2.

608336 Brownian Dynamics Simulations of Colloidal Sheets in Shear Flow

Abstract:

As 2D materials such as graphene, transition metal dichalcogenides, and 2D polymers become more prevalent, solution processing and colloidal-state properties are being exploited to create advanced and functional materials. However, our understanding of the fundamental behavior of 2D colloidal sheets and membranes in flow is still lacking. In this work, we perform Brownian dynamics simulations of semiflexible colloidal sheets with hydrodynamic interactions in shear flow. Different conformational behaviors, ranging from seemingly chaotic towel-wringing to periodic tumbling are observed as bending energy, stretching energy, and shear flow strength are varied. Parallels and differences with semiflexible polymer chains will be discussed.

606010 Investigating the Correlated Dynamics of Aqueous Solutions through Van Hove Functions: Insights from Molecular Simulation

Abstract:

The physical properties of water and aqueous solutions are of fundamental importance in innumerable fields. For several decades, multiple molecular simulation methods have guided experimental advances by describing the microscopic behavior of molecules and how such interactions affect macroscopic properties. As a result, it is of continued interest to evaluate the accuracy of various molecular simulation methods in describing these macroscopic behaviors in detail.

To fully understand the relationship between microscopic dynamics and macroscopic properties, such as viscosity, the collective and correlated dynamics of water must be evaluated. However, atomic correlations of fluids are often studied as static structures computed via ensemble averages. Recent advances in neutron and X-ray scattering experiments[1] have made it feasible to study the Van Hove correlation function of water and aqueous solutions[2]. These advances have enabled the study of the time-dependent structure of these fluids at very fine resolutions, on the order of less than a picosecond in time and less than an angstrom in distance. The signal produced from these experiments however, is a statistical average over all atoms of a system, making it difficult to determine which atomic interactions most strongly drive molecular behavior. Therefore, molecular simulation is an invaluable method alongside scattering experiments to evaluate the time-dependent behavior, as the total Van Hove correlation can be discretized into separate atomic components[3]. Here we present Van Hove correlation results aimed at better understanding the robustness of different force fields and molecular simulation techniques describing water and aqueous solutions. Evaluated and compared to experiment in this study are the Van Hove correlation functions computed from a range of classical, polarizable, and reactive force fields, as well as *ab initio* molecular dynamics simulations and density functional tight binding.

[1] Iwashita, T., Wu, B., Chen, W.-R., Tsutsui, S., Baron, A. Q. R., & Egami, T. (2017). Seeing real-space dynamics of liquid water through inelastic x-ray scattering. *Science Advances*, *3*(12), e1603079. https://doi.org/10.1126/sciadv.1603079

[2] Van Hove, L. (1954). Correlations in space and time and born approximation scattering in systems of interacting particles. *Physical Review*, *95*(1), 249–262. https://doi.org/10.1103/PhysRev.95.249

[3] Shinohara, Y., Matsumoto, R., Thompson, M., Dmowski, W., Ryu, C. W., Iwashita, T., Ishikawa, D., Baron, A., Cummings, P., Egami, T. (2019). Identifying Water-Anion Correlated Motions in Aqueous Solutions through Van Hove Functions. *Journal of Physical Chemistry Letters,* 10(22), 7119-7125. https://doi.org/10.1021/acs.jpclett.9b02891

607906 Structure/Dynamics of Electrolyte at Various Graphitic Electrode Interfaces

Abstract:

Commercial electrolytes for lithium ion batteries consist of ethylene carbonate (EC), diethyl carbonate (DEC), dimethyl carbonate (DMC), and lithium salt ($LiPF_6$); some contain low concentrations of additives such as fluoroethylene carbonate (FEC), propylene carbonate (PC), and vinylene carbonate (VC). Solvents and additives have been shown to strongly influence the performance of electrodes by altering the SEI layer composition, thickness, and conductivity.[1] The addition of additives has consistently been demonstrated as advantages, sometimes necessary for better performance. However, still much needed to be understood, cause certain surface chemistry/binder pairing shown to be removing the positive effects of FEC additives completely.[2]

Due to complexity of chemical and electrochemical reactions involved, as well as lack of experimental tools for the direct measurement, it is still difficult for current experimental methods to characterize the SEI (beyond chemical composition). This leads to a deficit in the understanding of the thermodynamic and kinetic phenomena associated with the process. Molecular simulations, however, grant a level of detail that is not afforded to experimentalists. Understanding the structure and dynamics of electrolyte at the electrolyte-electrode interface is essential in understanding reaction mechanisms for SEI formation. Here, we investigate the interplay between surface characteristics and interfacial structure and dynamics for concentrated electrolytes with lithium salt over four different interfaces (hydroxyl-, hydrogen-, carbonyl-, and carboxyl-terminated) using classical force fields with molecular dynamic simulations.

1. Zhang, S. S. A review on electrolyte additives for lithium-ion batteries. *Journal of Power Sources* **162**, 1379–1394 (2006).
2. Pandres, E. P., Olson, J. Z., Schlenker, C. W. & Holmberg, V. C. Germanium Nanowire Battery Electrodes with Engineered Surface-Binder Interactions Exhibit Improved Cycle Life and High Energy Density without Fluorinated Additives. *ACS Appl. Energy Mater.* **2**, 6200–6208 (2019).

596153 Utilizing Molecular Dynamics Simulations to Predict the Modulation of Interfacial Hydrophobicity By Chemical and Physical Properties

Abstract:

Predicting hydrophobic interactions is a complex challenge because of the cooperative and competing effects on hydrophobicity that arise from the interfacial chemical and physical properties of functionalized materials. The current understanding of hydrophobicity at the nanoscale relates hydrophobic interactions, surface properties, and the structure of interfacial water molecules. Developing experimentally validated simulation models that accurately capture these relationships is necessary to predict how surface properties modulate hydrophobic interactions. This type of predictive framework would be valuable for the design of materials that control processes driven by hydrophobic interactions, such as amphiphile self-assembly, nano-bio interactions, and peptide-surface binding.

In this study, we utilized classical molecular dynamics (MD) simulations with enhanced sampling techniques to explain the experimentally observed modulation of interfacial hydrophobicity by the physical and chemical properties of self-assembled monolayers (SAMs). First, we used this methodology to explain how the molecular-level order of uniformly nonpolar SAM alters hydrophobic interactions by perturbing interfacial water structure. Next, we applied this methodology to SAMs with polar and nonpolar groups mixed on the surface (chemically heterogeneous SAMs) to understand how specific chemical groups modulate hydrophobicity. This study revealed how each group uniquely perturbs the interfacial water structure and hydrogen bonding network, resulting in changes to the hydrophobicity of the SAMs. Finally, using the insights gained from the previous two studies, we modeled small (< 5 nm radius) SAM-protected gold nanoparticles (AuNP) to understand how both physical and chemical heterogeneities cooperatively and/or competitively influence hydrophobicity. The physical heterogeneity arises from the curvature of the AuNP whereas the chemical heterogeneity arises from the different chemical groups decorating the surface of the AuNP. We developed an analysis technique to map hydrophobicity onto the complex geometries of the AuNPs to understand how these features perturb the interfacial water structure and hydrophobicity. The results from our study provide new insights into the relationship between interfacial chemical and physical properties and hydrophobicity that can be used to guide the design of functionalized materials.

605239 Understanding the Homogeneous Ice Nucleation Process for Different Ice Polymorphs

Abstract:

Water is perhaps the most ubiquitous molecule on Earth and because of its features, it may exhibit various polymorphic structures each of which strongly depend on the thermodynamic conditions applied during nucleation. However, nucleation is complicated to observe both in experiment and computational work as it is considered to be a rare event. To undergo nucleation, a liquid is cooled down below its melting point reaching a metastable state. Using molecular dynamics simulations, we navigate the phase diagram of the TIP4P/Ice model to improve our understanding of the nucleation process under different temperatures and pressures. Moreover, molecular dynamics allow for the analysis of local order using parameters, such as the tetrahedral order parameter, to characterize polymorphic transformations (hexagonal ice and ices II through V). Because homogeneous ice nucleation often is a result of the competition between local density fluctuations and hydrogen bond formations we also monitor these features throughout the simulations.

595965 High-Throughput Approach to Studying Adsorption on Zeolite Surfaces

Abstract:

Zeolites are nowadays abundantly used as adsorbents and catalysts for many industrial applications. Computational studies have recently complemented the widely available experimental data and observations on zeolites by shedding light on reaction mechanisms and the active sites. The computational studies of zeolites, however, are challenging due to the numerous possible active sites. A zeolite structure can have many unique tetrahedral sites (T). Substitution of a trivalent aluminum atom to a tetravalent silicon atom in the zeolite framework leads to a net negative charge that can be compensated by a proton or a cation. However, there are numerous possibilities for the placement of Al and the compensating proton or cation. The number of possibilities grow exponentially for higher Al/Si ratios in the framework. Due to this complexity, most theoretical studies have focused on a limited number of T sites based on physical intuitions. For example, many studies of the MFI zeolite, which has 12 T-sites, have focused only on one or two T sites based on the accessibility of guest molecule to be the active site. These choices are justifiable given the high computational cost of accessing the full descriptor space, however, they probably miss promising active sites. In this work, we attempt to thoroughly explore the possible active sites using a high-throughput approach. We apply our approach to the problem of NO adsorption on metal-exchanged zeolites. It is known that Pd-exchange CHA, for example, is an effective adsorbent to limit NOx emissions. Full exploration of this problem requires enumerating the possible Al-Al pairs on the starting zeolites, identifying the favorable sites for the compensating protons and cations and searching for a global minimum in each case, which means thousands of calculations even for simple zeolites. The problem is further compounded when the cation is a metal that might have multiple oxidation states. We show how we systematically explore the possible sites, starting with a lower level of theory on the initial calculations, and then narrowing the potential candidates and using a higher level of theory afterwards. The automation of this calculation reduces the amount of manual labor and highly improve the efficiency of the workflow. It presents a new approach to tackle the complicated problems related to zeolite catalysis.

607612 Machine Learning-Guided Flow Synthesis of Inorganic Metal Halide Perovskite Quantum Dots

Abstract:

The controlled synthesis of advanced colloidal nanomaterials is limited by the currently available reactors utilized for their synthesis and the disjointed workflow between experimentation and data processing. Manual flask-based experimentation strategies, although applied across most research laboratories, are unlikely to sufficiently capture and control the complex and highly sensitive reaction systems of many colloidal nanomaterial syntheses. These batch reactors are slow, labor-intensive, and subject to batch-to-batch variations in experimental data sets. In response, over the last decade different microfluidic synthesis strategies have been developed and effectively applied towards the controlled synthesis of different classes of colloidal nanomaterials. Such flow synthesis strategies notably enhance the efficiency of colloidal synthesis studies, while simultaneously unlocking previously unattainable precision in synthesis parameter control. Despite the effectiveness of flow synthesis microreactors in accelerating reaction condition screening of colloidal nanomaterials, they still rely on user-directed experiment selection. Broad reaction space screening using design of experiments is materially inefficient, and expertly guided experiments are often too slow to effectively cover a synthesis space within the lifetime of unstable chemical precursors. These bottlenecks can be addressed through development of self-optimizing fluidic reactors and integrating fully automated experimentation, materials diagnostics, and data analysis with artificial intelligence-guided experiment-selection algorithms. Such autonomous materials synthesis platforms will be able to isolate optimal reaction parameters entirely without user intervention.

Colloidal semiconductor nanocrystals (known as quantum dots) are a particularly suitable testbed for autonomous materials development strategies. Quantum dot syntheses possess a large and highly sensitive reaction parameter space.[1] Proper control of their reaction conditions is difficult to attain both within a single set of precursors and across research labs (batch to batch variation). As a result, despite extensive efforts and constantly growing applications, many synthesis paths are either unexplored or not fully optimized. One of the recently emerging classes of quantum dots is colloidal metal halide perovskites, which have highly favorable properties for applications in optoelectronic devices. Due to the combinatorially increasing synthesis space of metal halide perovskite quantum dots, many of their possible synthesis routes are novel and unexplored. Therefore, these reaction systems present an exciting opportunity for exploration within an autonomous flow synthesis platform.

In this work, we present the first machine learning guided-fluidic reactor for the controlled synthesis of high-quality colloidal quantum dots. While the strategies and techniques detailed here may be easily applied to a large number of solution-processed materials, as a case study we analyzed the halide exchange of cesium-lead-bromide perovskite quantum dots with zinc halide salts and two different surface ligands.[2] We developed a novel *in situ* material diagnostic module to attain accurate real-time access to peak emission energy, emission linewidth, and photoluminescence quantum yield of the in-flow synthesized perovskite quantum dots. These output parameters, corresponding to the physicochemical and optoelectronic properties of quantum dots, were then simultaneously optimized through an objective function integrated within a custom neural network ensemble model and various decision-making policies (used to select conditions from the model). Utilizing the developed autonomous materials development platform, we evaluated the performance of eight different experiment selection algorithms (spanning machine learning to evolutionary algorithms) over the course of more than 1400 experiments. The controlled syntheses of perovskite quantum dots were performed without user intervention and required less than 400 μL of starting quantum dot solution per experimental condition. The methods detailed in this work offer autonomous reaction optimization across multiple vital performance parameters. Further use of similar self-optimizing systems in the field of colloidal nanoscience would reduce the cost and expedite the process of materials development and discovery.

References

[1] R. W. Epps, K. C. Felton, C. W. Coley, M. Abolhasani, *Lab Chip* 2017, *17*, 4040.

[2] K. Abdel-Latif, R. W. Epps, C. B. Kerr, C. M. Papa, F. N. Castellano, M. Abolhasani, *Adv. Funct. Mater.* 2019, *29*, 1900712.

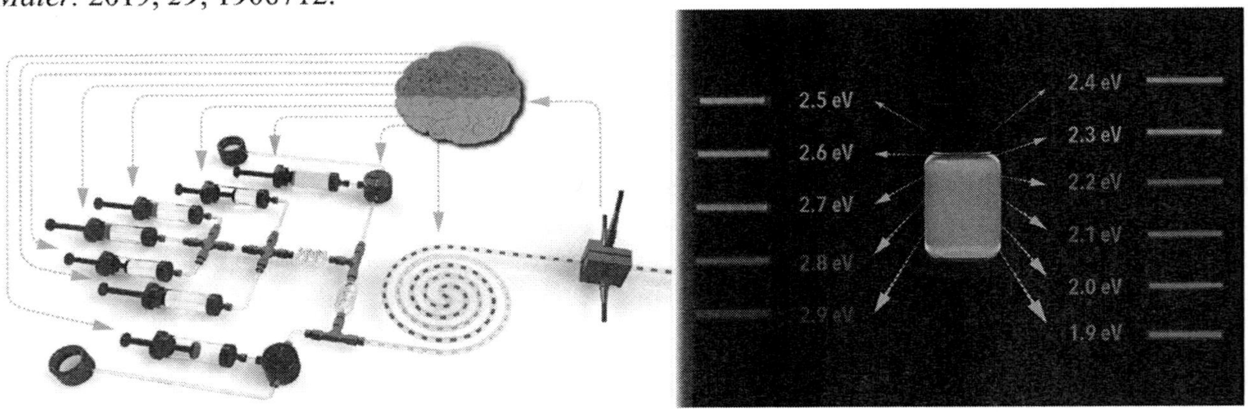

609449 High-Throughput and Data-Driven Strategies for the Design of Deep Eutectic Solvent Electrolytes

Abstract:

Within the framework of green chemistry, Deep Eutectic Solvents (DES) have been identified as promising candidates for use in many applications, including battery electrolytes. DES are characterized by two or three materials that associate with each other through hydrogen bond interactions, resulting in a eutectic mixture whose freezing point is below that of the individual materials. This design space is overwhelmingly large and poses a challenge for screening a vast and diverse set of materials. Here we present a strategic approach consisting of high throughput experimentation (HTE) coupled with data science driven analysis to identify exceptional DES candidates based on key physiochemical and electrochemical properties. Much of our HTE adopts methods that are already used frequently in the biotech and pharmaceutical industries, most notably performing parallel syntheses and analyses in 96-well-plate formats. DES samples are first synthesized using an open-sourced automated liquid handling robot. DES melting points are then determined by monitoring the melting process with an infrared camera and identifying the temperature at which the thermal conductivity of the samples changes abruptly. The solubility of battery redox-species is determined via UV-VIS well-spectrophotometers. Finally, the electrochemical stability window and cycling properties of DES electrolytes are measured in high-throughput by using screen-printed electrodes on 96-well plates adapted for use with a standard potentiostat. The ability to rapidly and efficiently collect data also creates a need for the development and use of automated processes for data analysis, which have been developed in an open-sourced format by our group. This approach to HTE also allows for the incorporation of data science techniques, such as feature extraction and machine learning, that further aid in probing a design space that is ultimately too large for experimental methods alone.

602248 Development of a Hybrid Model to Describe the Dynamics of a Partially Known System

Abstract:

Mathematical modeling plays a critical role in the design, optimization, and control of chemical processes [1]. In order to apply models to a wide range of process conditions, it is often desirable to describe the system using a first principles model. However, the formulation of such models can be challenging if the underlying mechanisms are only partially understood [2, 3, 4]. Under such circumstances, a data-driven model can be built from measured data, and the resultant model can describe the system adequately, even when mechanistic understanding is limited [5, 6]. However, a data-driven model has narrow applicability as it is tailored to describe the input-output relationships contained in the training datasets [7]. As an alternative, a hybrid modeling approach that combines first-principles and data-driven modeling has been proposed to describe a partially known process [7, 8]. Hybrid models have better prediction capabilities than first-principle models, and thus will have better generalizability and interpretability than data-driven models.

In this work, we proposed a new numerical scheme to construct a hybrid model for a system that has partial mechanistic information. A hybrid model was constructed by introducing correction terms for estimating process-model mismatches between mechanistic model predictions and experimental measurements [9]. Here, a nominal model refers to a mechanistic model based on our prior knowledge of the system of interest, and the correction terms are assigned to differential equations of the nominal model so that the resultant hybrid model has improved prediction capabilities. The proposed methodology consists of three steps. First, observability analysis was performed for the nominal model to determine a subset of states that are observable from the available measurements. This step is critical to reduce the computational load of the estimation and avoid potential overfitting issues. Second, the available measurements were used to estimate the correction terms associated with the observable states through solving a L2-regularized least-squares problem [10]. Lastly, once the correction terms were estimated, the trajectories of the states were simulated based on the hybrid model. Then, an artificial neural network (ANN) model was developed for predicting the correction terms for a given set of state values, and the developed ANN was used within the hybrid model. Through incorporating the ANN, the hybrid model was able to predict dynamic responses to an input, which was not in the measurement data used for estimating the correction terms.

The validity of the proposed method was demonstrated by constructing a hybrid model for the NFκB signaling pathway induced by both lipopolysaccharide (LPS) and brefeldin A (BFA). The signaling dynamics induced by LPS have been well studied, and hence, a previously published model was used as the nominal (mechanistic) model [11]. Since the dynamics initiated by BFA are less characterized, the nominal model could not capture the signaling dynamics induced by both BFA and LPS well. By implementing the proposed methodology, the developed hybrid model was able to predict the NFκB signaling dynamics induced by both LPS and BFA accurately.

References

[1] Hangos, K. M.; Cameron, I.T. Process modelling and model analysis. Academic Press: London, 2001.

[2] Cho, K.-H.; Choo, S.-M.; Jung, S.H.; Kim, J.-R.; Choi, H.-S.; Kim, J. Reverse engineering of gene

regulatory networks, IET Systems Biology 2007, 1, 149 - 163.

[3] Willis, M.J.; von Stosch, M. Inference of chemical reaction networks using mixed integer linear programming, Computers and Chemical Engineering 2016, 90, 31-43.

[4] Lee, D.; Jayaraman, A.; Kwon, J.S. Identification of a time-varying intracellular signalling model through data clustering and parameter selection: application to NF-κB signalling pathway induced by LPS in the presence of BFA, IET Systems Biology 2019, 13, 169-179.

[5] Thompson, M.L.; Kramer, M.A. Modeling chemical processes using prior knowledge and neural networks, AIChE Journal 1994, 40, 1328-1340

[6] Narasingam,A.; Kwon, J.S. Development of local dynamic mode decomposition with control: Application to model predictive control of hydraulic fracturing, Computers & Chemical Engineering 2017, 106, 501-511.

[7] von Stosch, M.; Oliveira, R.; Joana Peres, J.; de Azevedo, S.F. Hybrid semi-parametric modeling in process systems engineering: Past, present and future. *Computers & Chemical Engineering* **2014**, *60*, 86-101.

[8] Faizan Bangi, M.S.; Kwon, J.S. Deep hybrid modeling of chemical process: Application to hydraulic fracturing, Computers & Chemical Engineering 2020, 134, 106696.

[9] Engelhardt, B.; Fröhlich H.; Kschischo, M. Learning (from) the errors of a systems biology model, Scientific Reports 2016, 6, 20772.

[10] Howsmon, D.P.; Hahn, J. Regularization Techniques to Overcome Overparameterization of Complex Biochemical Reaction Networks, IEEE Life Sciences Letters 2016, 2, 31 – 34.

[11] Lee, D.; Ding, Y.; Jayaraman, A.; Kwon, J.S. Mathematical modeling and parameter estimation of intracellular signaling pathway: application to LPS-induced NFκB activation and TNFα production in macrophages, Processes 2018, 6, 21.

607279 Ffiber: Framework for Fluorescent Neuroimaging Based Experimental Routines

Abstract:

Background: The modern lab produces more data than it can analyze using traditional experimental techniques. For neuroimaging labs, terabytes of imaging data can easily be generated in the course of a single experiment. However, only a small portion of this data is accessible for extracting results based on existing methods. To enhance the portion of data that can be analyzed, many labs are turning to data science for high throughput processing and analysis. However, to synthesize highly skilled domain scientists with an effective data science strategy, a robust method for developing well-designed and data informed experimental pipelines must be followed. We developed a framework for fluorescent neuroimaging based experimental routines (FFIBER) that implements key practices of data science, software engineering, chemical engineering, and neuroimaging. The goal of this pipeline is to produce experimental workflows for high throughput neuroimage acquisition and analysis. For each experiment we follow a six-step process: (1) develop a data awareness, (2) design a data management plan, (3) determine an optimal experimental pipeline, (4) build out supporting data science infrastructure, (5) perform primary and supplemental imaging, and (6) produce interpretable visualizations of results. The *objective* of this project is to apply our FFIBER method during initial experimental stages and retroactively to generate, process, and analyze high volumes of neuroimaging data without an overall increase in labor-hours.

Methods: To develop a data awareness, we performed a data assessment of relevant metadata, data storage, and data relationships. When designing a data management plan, we identified the optimal data storage locations, file structure for raw data and results, and personnel responsibilities. To determine an optimal experimental pipeline, we performed software reviews, addressed labor-hour bottlenecks, and identified necessary programs, packages, and scripts. Then we built out the supporting data science infrastructure utilizing Python and Jupyter notebooks as the base script for pipelines, electronic laboratory notebook integration, and version control with the team through GitHub and Google Drive. After the pipeline was built, we completed the primary set of experimental imaging using confocal microscopy of brain slices and fed the data through the pipeline for analysis. We then completed supplemental imaging to experimentally support a representative set of results obtained through FFIBER. Finally, we wrote Python scripts to output consistent interpretable visualizations of our analyses and align these with other experimental results.

Results: We applied FFIBER to a data science experiment analyzing the phenotype of microglial cells in *ex vivo* brain slices to assess extent of glial cell activation. We were able to create and modify a pipeline during the initial stages of (1) an oxygen-glucose deprivation experiment and retroactively for (2) an inflammation-sensitized model using *E. coli derived* lipopolysaccharide, (3) and a neonatal hypoxia-ischemia (HI) model. A preliminary data assessment revealed a storage necessity of 5 GB per experiment with five individual file types from five separate personnel. We completed the data assessment and built a data management plan with university-supported Google Drive as our storage location, *.tiff*, *.csv,* and *.ipynb* as our file types, and established a slice-based, double-blind file structure. We wrote Python scripts automating image upload, cleaning and automatic feed to our analysis pipeline that reduced labor-hours from five hours per image set to less than ten minutes. Our software review led us to integrate a published cell morphometrics package, VAMPIRE, with our diff_register package for skeletonized morphometrics of microglia. Our image analysis pipeline developed with FFIBER split uploaded confocal images into fourths, segmented, skeletonized and identified the microglia in the images, and uploaded the images into our integrated morphometric analysis. FFIBER structured the development of our work pipeline that decreased work hours from a full day to less than an hour to analyze an entire experiment. Finally, our FFIBER structured pipeline enabled us to detect and quantify shape differences between non-treated and injured *ex vivo* slices – providing additional information for our initial oxygen-glucose deprivation experiment and retroactive insight to our inflammation sensitized and neonatal HI models. Our first application of FFIBER decreased our human labor time from > 24 hours per 5 GB image set to less than an hour, decreased our storage budget to $0, and produced information about cell morphometrics previously unavailable to our lab due to lack of analysis software and time constraints.

Conclusions: By applying FFIBER, we were able to create workflow pipelines both during initial stages of experimental design and retroactively in neuroimage sets already collected. We decreased labor hours for data sets from 24 hours to less than an hour through using Python scripts rather than manual work and a well-designed pipeline. Our framework creates a high throughput approach for experimental design in research dependent on neuroimaging by integrating traditional experimental methods with modern data science applications. We can further iterate the methodology to include metadatabases for easier data access and data connections, budget vs storage matrices for determining best data storage location, and templates for electronic lab notebooks, file structures, data relationships, and generalized processing pipelines. Additionally, while FFIBER was developed for fluorescent neuroimaging datasets, the current workflow is robust, and flexible enough for application to other image analysis techniques, other organs, and other fluorescent imaging methods.

597491 Data-Driven Discovery of Polymeric Vehicles for Gene Editing:Serendipity-Inspired Design Directions.

Abstract:

Synthetic materials such as cationic polymers are poised to replace engineered viruses as vehicles for therapeutic nucleic acids. Polymeric materials offer a vast design space where desired material properties can be engineered by deft optimization of chemical composition and polymer architecture. However, rational polymer design is impeded by the "curse of dimensionality" since systematic examination of numerous variables such as composition, architecture, length and formulation parameters is paramount. To accelerate polymeric vector discovery, we need to develop models that predict the impact of key design attributes such as phase behavior, basicity, surface charge, nucleic acid binding, and hydrophobicity on the ultimate biological responses observed. In this work, we eschew low-throughput hypothesis-driven design of polymers in favor of a data-driven approach that exploits combinatorial polymer libraries and high-throughput experimentation. Thereafter, we employ statistical learning to extract the molecular attributes of the hit polymers that promoted gene editing efficiency and apply this knowledge to provide experimental guidance on designing subsequent polymer libraries.

Drawing inspiration from high-throughput experimental approaches used in the pharmaceutical industry to screen drug libraries for therapeutic activity, we employed parallel polymer synthesis, formulation and well plate based biological assays for rapid screening of gene editing efficiency We synthesized a chemically diverse library of copolymers combining different ratios of 1) cationic monomers bearing amines spanning a broad range of basicity and 2) neutral monomers of varying hydrophilicity. Our combinatorially designed library allows for systematic investigation of the effect of amine basicity by studying variations in polymer pKa resulting from the use of 4 cationic monomers, in ratios of 100, 75, 50 and 25 %. Subsequent to parallel polymer synthesis, extensive physicochemical characterization was completed using automated tools -: composition and molecular weight analysis, pK_a, polyplex size distribution, binding assays, and ζ-potential measurements were acquired in high-throughput modes to generate a rich dataset. Polymers were complexed with ribonucleoprotein (RNP) payloads and gene editing was quantified by estimating the proportion of mCherry positive cells. To resolve the trade-off between sensitivity and experimental throughput, we employed image cytometry and developed an image processing algorithm to quantify mCherry expression in a robust and automated fashion from a bank of images arising from 200 unique formulations. At the end of the high-throughput screening campaign, we obtained a high-performing hit polymer that outperformed state-of-the-art synthetic transfection reagents.

Having identified a hit formulation from our library, the challenge was to: 1) unravel the relationship between polymer attributes and gene editing efficiency. 2) build on these structure-activity relationships to guide the design of future polymer libraries that will yield a higher "hit rate". In order to derive these predictive relationships, we sought to understand how 10 polymer descriptors influenced biological performance. We turned to principal component analysis (PCA), to deal with the dimensionality challenge created by our dataset. Through PCA, we concluded that a single polymer descriptor cannot be used in isolation to guide the design of future libraries. Rather, complex non-linear relationships between several molecular attributes were responsible for editing performance. Discarding preconceived notions of how various chemical functional groups will influence delivery, we screened a large chemically diverse polymer library to discover design guidelines that do not conform to traditional heuristics as well as a promising hit polymer, which may not have been accessible through hypothesis-testing. If statistical learning and automated experimental workflows are applied in tandem, we can overcome challenges originating from the complexity of structure-function relationships governing polymeric gene delivery.

605093 Artificial Intelligence-Assisted Full-Map Understanding of Strain Sensing Devices and Designated Predictions for Soft Robotic Systems

Abstract:

Compliant strain sensing devices are widely applied in real-time mechanical feedback for soft robotic systems. However, it remains challenging to reach customized sensing devices with high sensitivity coordinated with designated working windows due to the complex and unclear structure-composition-property relationships which are hard to predict and pre-design. In this work, through active learning and developed algorithms, we propose machine assisted full-map structure-composition-property understandings of strain sensors consist of classical nanocomponents of $Ti_3C_2T_x$ MXene, carbon nanotube, and poly(vinyl alcohol). Important variables of nanocomponent weight loading, nanocoating thickness, and nanocoating surface morphology are systematically investigated, and an informative data base (5000+ experimental resistance points, 500+ nanocoatings, and 200+ sensing devices) is constructed for training of machine learning models. Besides, comprehensive algorithm optimizations are performed for better machine performance. Ultimately, the machine learning models show low average scattered numerical error of 24% and give customized predictions in 1) sensing performance in a given material recipe; 2) suitable material recipes of sensing devices with high sensitivity and designated working windows in soft robotic systems. The reported strategies here open new perspectives towards design-free high-performance strain sensing devices and customized applications in soft robotic systems.

More specifically, in our material systems, at least 33,700 non-repeating recipes are available for strain sensing devices fabrication although we limit the step length in each variables of material recipes. With such a complex system and very limited understanding, there are no available *in silico* simulation methods to interpret or predict the sensing devices. In addition, it is nearly impossible to screen the whole design space *via* Edison approach because the sensing needs are customized every time. Herein, we have proposed a data-driven framework to exploratively learn the full-map of the strain sensing devices. The data is collected with thirteen active learning rounds which investigate both the mean minimal distance (L_2) between suggestion points and existing data points, and the numerical uncertainty of well-trained models during each round. After a representative dataset is constructed, optimization strategies including GA combination, customized loss function, and data augmentation are used for improving the model performance. This best model has achieved a low average scattered numerical error ((Predicted value - Real value)/Real maximum value in device) of 24%.

608059 An Autonomous Computational Workflow for Efficient Generation of Polymer Data

Abstract:

Data development is critically important for the emerging polymer informatics ecosystems. Two primary challenges of preparing a computational polymers database are which polymer (out of a huge number of possible candidates) should be considered and how to obtain a reasonable polymer model, from which calculations for relevant can be performed. We have developed an autonomous computational workflow, involving an active-learning guidance step and an efficient Polymer Structure Prediction strategy, to solve these problems. Our scheme has been implemented in the Computational Autonomy for Materials Discovery package and used to significantly enlarge the computational polymer database that power the Polymer Genome platform (https://www.polymergenome.org/). Machine-learning models trained on our computational data for predicting various polymer properties are available in https://www.polymergenome.org/. This work is supported by Toyota Research Institute and Office of Naval Research.

597874 Leveraging Experimental Transition Metal Complex Information to Improve Generalizability of Machine Learning Models

Abstract:

Creating computationally inexpensive, yet accurate, models for materials and molecular properties is a primary driver for data science in chemical engineering. However, extending these models to target systems beyond the training data often requires models to aggregate large quantities of information from different chemistries. Our group has previously generated accurate artificial neural network (ANN) models trained on density functional theory (DFT) properties for mononuclear octahedral transition metal (TM) complexes primarily through exhaustive enumeration. Here, towards expanding the applicability of our ANN models to more diverse chemical space, we first mined all experimental, structurally identified mononuclear octahedral TM complexes in the Cambridge Structural Database (CSD). By featurizing these complexes with revised autocorrelation functions, a class of graph-based heuristic descriptors developed in our group, we develop a bird's eye view of their similarity to each other and to prior data generated in our group's database. Next, we compare a number of approaches for selecting the next CSD complexes to evaluate by DFT, with the aim of accelerating the generalizability of re-trained ANN models to these diverse chemistries. Approaches include physically-motivated sampling, such as most-diverse sampling of ligand symmetries, and ANN-derived uncertainty sampling, i.e., selecting complexes with the lowest model confidence. We anticipate our benchmarked methods for selecting complexes for evaluation will be useful across domains where rapid expansion to unseen materials is desired.

605838 A Formal Grammar-Based Machine Learning Approach for Predicting Reaction Outcomes

Abstract:

Prediction of reaction outcomes without performing time-consuming experiments offers several invaluable advantages -- reducing experimentation costs due to elimination of highly improbable reactions, lowering the time to hypothesis-validation-and-correction, enabling high-throughput experimentation for complex reactions, and allowing significantly more time for analyzing the results. In recent years, machine learning techniques have been successfully applied in this area that, either directly or indirectly, address one or more of the above issues. A few areas of application include discovery of novel molecules with desired properties, understanding structure-property relationships, identifying catalysts for better reaction catalysis, and data-driven modeling of chemical reaction kinetics. The success of machine learning methods could largely be attributed to their inherent proficiency in capturing complex non-linear dependencies between various factors that govern the reaction systems. Here, we present an approach for predicting the most likely product of a reaction based on a given set of reactants and agents (catalysts, reaction medium). The structural properties of reactants and the product molecules are encoded based on a formal grammar, akin to context-free grammars (CFG) in the area of natural language processing (NLP), that essentially gives rise to a parse-tree representation of the molecule. This representation is used in a convolutional neural network architecture to model the structural transformations from a set of reactants to the most likely product of the reaction. The proposed framework could be used for rapid experimentation and analysis of reaction outcomes, especially if the molecules involved are less-known and structurally novel. In addition, the latent space of the neural network architecture could also be used to discover insights on the transformations from reactants to products in a given reaction under different conditions.

611660 Inverse Learning of Material Physics through Image Data and Continuum Modeling

Abstract:

The advent of big data in spatio-temporal microscopic images as well as large-scale atomistic simulations offers a tremendous amount of hidden information about the physical chemistry at the continuum level. Using a framework of PDE-constrained optimization, we extract multiple unknown constitutive relations simultaneously from a small set of images of pattern formation. Compared to the data-driven modeling approach, our approach provides clear physical interpretability by prescribing a general governing equation while faithfully achieving quantitative matching between data and continuum model prediction through highly expressive nonlinear and/or nonlocal (integro-differential) constitutive relations.

We present examples of learning state-dependent properties such as the diffusivity, kinetic prefactor, free energy, and direct correlation function from Cahn-Hilliard equation, Allen-Cahn equation, and Phase Field Crystal Model. Applying the method to experimental images of lithium concentration field in lithium-iron phosphate (LFP) particles during charge and discharge, we are able to extract its free energy and reaction kinetics, and "image" its reaction kinetics heterogeneity map. In addition, we also demonstrate its ability to coarse-grain molecular simulations and hence provide acceleration by learning functions of free energy and diffusivity and achieving quantitative matching of the patterns of spinodal decomposition.

607537 Improved Training of Graph-Embedding Based Neural Network Energy Functions for Catalysis

Abstract:

Complex reaction networks comprising of thousands of intermediates and reactions are ubiquitous in heterogeneous catalysis; developing mechanistic models of such systems is computationally intractable primarily due to the cost of computing the energies of reactions and species using an ab initio method such as density functional theory (DFT). Developing machine learned energy potentials trained on DFT energies offer a tractable way of building mechanistic models for these systems. To this end, the contribution of this talk is two-fold. First, we develop a novel deep neural network model to learn and predict the potential energy surface (PES) for periodic systems. The novel architecture integrates graph embedding of individual atoms, atomic interactions and periodic boundary conditions with physics-aware feature engineering, residual learning, and attention mechanism. Second, we train the model with an efficiently distributed Hessian-free optimization method and deliver a prediction within chemical accuracy, which demonstrates the potential of second order methods in the joint field of deep learning and quantum chemistry. We will use examples from the field of heterogeneous catalysis, including (i) alkane conversion on molybdenum sulfide and (ii) adsorption of hydrocarbons in zeolites.

Explainability based Fault Detection and Diagnosis using Deep Learning

Piyush Agarwal[1], Melih Tamer[2], and Hector Budman[1]

[1]Chemical Engineering Department, University of Waterloo, Waterloo, Canada, N2L3G1
[2]Sanofi Pasteur, Toronto, Ontario M2R 3T4, Canada

April, 2020

The fourth industrial revolution also known as 'Industry 4.0' and Big Data revelations has enabled the manufacturing industries to boost its performance in terms of operation, economic and safety. In order to meet high expectations of quality, efficiency and reliability of the product, a process monitoring system is needed. The two important aspects of Statistical Process Monitoring (SPM) are fault detection and diagnosis (FDD). Safe operation of process plants can be ensured by determining if the current state of the process is normal or abnormal. This problem is known as "Fault Detection". Following the detection of abnormality in the process, the next step is to classify the fault and diagnose its causes, also known as "Fault Diagnosis/ Classification". Noisy and high dimensional recorded process inputs greatly hinders the FDD mechanism in process plants. Previously, traditional linear PCA and its variants [5, 6] based methods have been extensively used for fault detection and prognosis. However, the inherent non-linearity in the process poses challenges for these methods. To this end, deep learning methods have shown considerable improvement over traditional methods [3, 4]. Deep learning methods can be broadly classified into two categories: 1) Unsupervised Learning; 2) Supervised Learning. In this work we use a supervised approach on a benchmark Tennessee Eastman Process (TEP) and show the improvement over both PCA-PLS based methods, unsupervised learning and previously used supervised learning based methods. This work is motivated by the fact that deep learning algorithms require generally large amounts of data for proper training and thus, their accuracy may be compromised when using relatively small datasets (bottleneck for many industries such as pharmaceutical industries etc.). Therefore, there is a large incentive to increase the accuracy of deep learning based FDD models without increasing the amount of data for training. Towards this goal the concept of explainability of Neural Networks (NNs) is exploited to improve the accuracy of the FDD without increasing the amount of data. In this study, the explainability of the effect of the inputs on the output classes is quantified by relevances interpreted using Layer-wise Relevance Propagation (LRP) and the irrelevant input variables' for the supervised classification problem are discarded thus enhancing both fault detection and fault diagnosis accuracies [1, 2]. This relevance based procedure greatly reduces the number of tuning parameters in the network and also provide physical insights for the classification problem at hand. Lastly, we use Dynamic Deep Supervised Autoencoder (DDSAE) NNs which improves the classification ability compared to traditionally used SAE (Stacked Autoencoder) NNs. The efficacy of the proposed method over different methods for both Fault Detection and Diagnosis are shown in Table 1 and Figure 1 respectively.

The proposed deep learning based detection method is compared with linear Principal Component Analysis (PCA), Dynamic Principal Component Analysis (DPCA) , Independent Component Analysis (ICA) and with two other recently reported methods that use Deep Learning models (architecture used: Sparse Stacked Autoencoder NNs (SAE-NN) and Convolutional NN (CNN)) for the same data set. For the Fault Diagnosis problem, the proposed method is compared with Support Vector Machines (SVM), Random Forest, Structure SVM, and sm-NLPCA (architecture used: Stacked Autoencoder). It is observed that the proposed relevance based method significantly increases the average fault detection accuracy over other methods.

References

[1] Piyush Agarwal and Hector Budman. Classification of profit-based operating regions for the tennessee eastman process using deep learning methods. *IFAC-PapersOnLine*, 52(1):556–561, 2019.

[2] Piyush Agarwal, Melih Tamer, M Hossein Sahraei, and Hector Budman. Deep learning for classification of profit-based operating regions in industrial processes. *Industrial & Engineering Chemistry Research*, 2019.

Table 1: Comparison of Fault Detection rate with different methods

Fault	PCA (15 components)		DPCA (22 components)	ICA (9 components)		DL (2017)	DL (2019)	Proposed DL Methodology
	T^2	SPE	T^2	I^2	AO	SAE-NN	CNN	DDSAE
1	99.2	99.8	99	100	100	77.6	91.39	99.9
2	98	98.6	98	98	98	85	87.96	98.2
4	4.4	96.2	26	61	84	56.6	99.73	99.7
5	22.5	25.4	36	100	100	76	90.35	100
6	98.9	100	100	100	100	82.8	91.5	99.7
7	91.5	100	100	99	100	80.6	91.55	99.9
8	96.6	97.6	98	97	97	83	82.95	94
10	33.4	34.1	55	78	82	75.3	70.05	89.1
11	20.6	64.4	48	52	70	75.9	60.16	87.7
12	97.1	97.5	99	99	100	83.3	85.56	99.4
13	94	95.5	94	94	95	83.3	46.92	95.6
14	84.2	100	100	100	100	77.8	88.88	100
16	16.6	24.5	49	71	78	78.3	66.84	94
17	74.1	89.2	82	89	94	78	77.11	97.7
18	88.7	89.9	90	90	90	83.3	82.74	90.9
19	0.4	12.7	3	69	80	67.7	70.87	89.9
20	29.9	45	53	87	91	77.1	72.88	89.6
Average	61.77	74.72	72.35	87.29	91.70	77.74	79.84	95.60

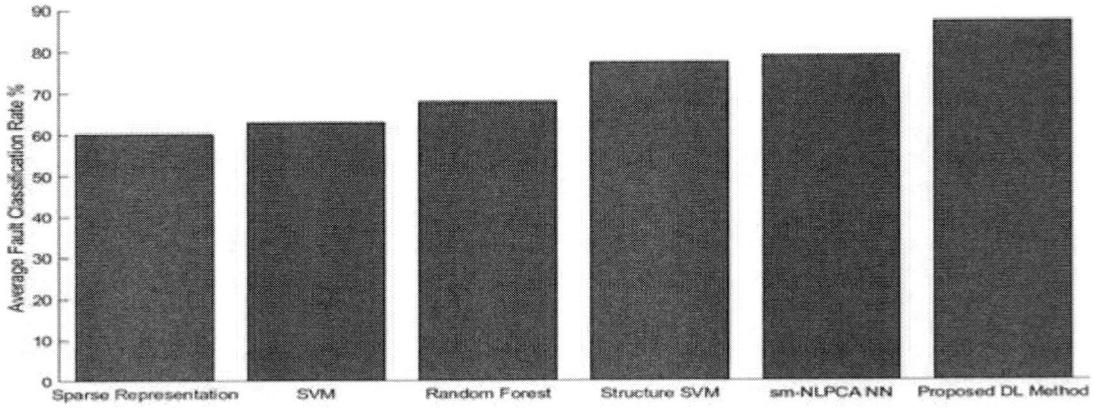

Figure 1: Comparison of Fault Classification rate with different methods

[3] Gavneet Singh Chadha and Andreas Schwung. Comparison of deep neural network architectures for fault detection in tennessee eastman process. In *2017 22nd IEEE International Conference on Emerging Technologies and Factory Automation (ETFA)*, pages 1–8. IEEE, 2017.

[4] Seongmin Heo and Jay H Lee. Statistical process monitoring of the tennessee eastman process using parallel autoassociative neural networks and a large dataset. *Processes*, 7(7):411, 2019.

[5] Barry M Wise, NL Ricker, DF Veltkamp, and Bruce R Kowalski. A theoretical basis for the use of principal component models for monitoring multivariate processes. *Process control and quality*, 1(1):41–51, 1990.

[6] Xiaofeng Yuan, Jiao Zhou, Yalin Wang, and Chunhua Yang. Multi-similarity measurement driven ensemble just-in-time learning for soft sensing of industrial processes. *Journal of Chemometrics*, 32(9):e3040, 2018.

601930 Physical Graph Neural Networks for Prediction of Fuel Ignition Quality

Abstract:

The increase in CO_2 emissions calls for the development of novel fuels that are based on renewable resources and can be used as a sustainable alternative in internal combustion engines (ICE). To assess the quality of a molecule as a fuel candidate, combustion-related properties need to be evaluated [1]. Fuel ignition quality indicators like the derived cetane number (DCN), the research octane number (RON), and the motor octane number (MON) give first, valuable insights on whether a molecule is suitable as an ICE fuel component. To predict these indicators and thus allow for facile molecular screening, many models based on quantitative structure-property relationships (QSPRs) have been proposed in the literature [2-5]. These models rely on manually selected descriptors to specify the molecular structure and apply regression methods to perform property prediction. The prediction quality, therefore, depends on the choice of informative descriptors.

Recently, graph neural networks (GNNs) [6,7] have shown promising results for the prediction of structure-property relationships of molecules [8]. GNNs use graph representations of molecules, where nodes correspond to atoms and edges to bonds. These graph representations are then linked to molecular properties by means of graph convolutions, pooling layers, and multilayer perceptron networks. More specifically, GNNs learn physico-chemical properties as a function of the molecular graph in a supervised learning set-up using a backpropagation algorithm. This end-to-end learning eliminates the need for selection of functional groups or descriptors. Rather, the method learns optimal molecular fingerprints through graph convolutions and maps the fingerprints to the physico-chemical properties by deep learning.

We propose a GNN model for predicting the DCN, MON, and RON of (oxygenated) hydrocarbons [9]. The model combines our higher order GNN with a recurrent neural network architecture. Further, it includes a physically meaningful pooling function which adds up contributions of individual atoms to the molecular descriptor. One of the main challenges in this work is the limited availability of experimental data. In order to overcome this difficulty, we study multitask learning, transfer learning, and ensemble learning. The results show competitive performance of the proposed GNN approach compared to state-of-the-art QSPRs making it a promising field for future research. The model and training code is open-source available at

https://git.rwth-aachen.de/avt.svt/public/graph_neural_network_for_fuel_ignition_quality
and a prediction tool is available via a web front-end at http://www.avt.rwth-aachen.de/gnn

[1] Gschwend, D., Soltic, P., Wokaun, A., & Vogel, F. (2019). Review and performance evaluation of fifty alternative liquid fuels for spark-ignition engines. Energy & Fuels, 33(3), 2186-2196.

[2] Dahmen, M., & Marquardt, W. (2015). A novel group contribution method for the prediction of the derived cetane number of oxygenated hydrocarbons. Energy & Fuels, 29(9), 5781-5801.

[3] Kubic Jr, W. L., Jenkins, R. W., Moore, C. M., Semelsberger, T. A., & Sutton, A. D. (2017). Artificial neural network based group contribution method for estimating cetane and octane numbers of hydrocarbons and oxygenated organic compounds. Industrial & Engineering Chemistry Research, 56(42), 12236-12245.

[4] Kessler, T., Sacia, E. R., Bell, A. T., & Mack, J. H. (2017). Artificial neural network based predictions of cetane number for furanic biofuel additives. Fuel, 206, 171-179.

[5] Smolenskii, E. A., Bavykin, V. M., Ryzhov, A. N., Slovokhotova, O. L., Chuvaeva, I. V., & Lapidus, A. L. (2008). Cetane numbers of hydrocarbons: calculations using optimal topological indices. Russian Chemical Bulletin, 57(3), 461-467.

[6] Kipf, T. N., & Welling, M. (2017). Semi-supervised classification with graph convolutional networks. Proceedings of the 5th International Conference on Learning Representations.

[7] Hamilton, W., Ying, R., & Leskovec J. (2017). Inductive Representation Learning on Large Graphs. Proceedings of the 30th Annual Conference on Neural Information Processing Systems, 1024-1034.

[8] Coley, C. W., Jin, W., Rogers, L., Jamison, T. F., Jaakkola, T. S., Green, W. H., Barzilay, R. & Jensen, K. F. (2019). A graph-convolutional neural network model for the prediction of chemical reactivity. Chemical Science, 10(2), 370-377.

[9] Schweidtmann, A. M., Rittig, J. G., König, A., Grohe, M., Mitsos, A., & Dahmen, M. (2020). Graph Neural Networks for Prediction of Fuel Ignition Quality. Submitted for Journal Publication.

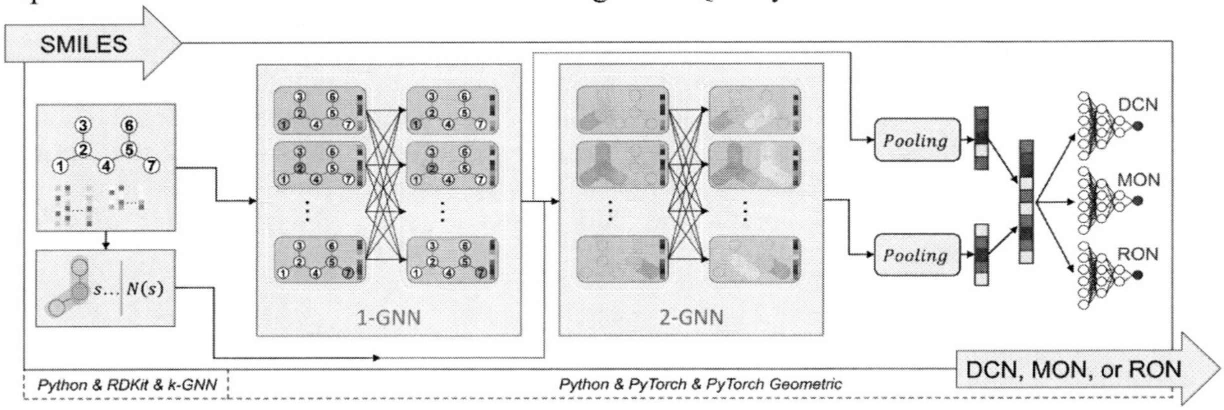

605089 Adaptive Spectral Graph Convolutional Neural Network in Crystal Property Prediction

Abstract:

Machine Learning (ML) has been more and more widely used in accelerating the design of new materials due to its close accuracy and less computational cost in predicting material properties comparing to simulation methods. However, existing ML methods for designing of crystalline materials are usually not flexible to handle different crystal types or hard to interpret because of the arbitrary size. Although there are more generalized models such as Crystal Graph Convolutional Neural Networks (CGCNN) which utilized graph CNN to encode the structure of a crystal, they are over-localized and impossible to learn beyond the bond connectivity. Here we show a stronger and more adaptive framework, based on spectral convolutional neural network, which is not only capable for arbitrary crystal structure but also learns a unique graph representation based on both global and local topology of each crystal.

First a crystal graph is produced base on the cif file of the crystal similar to CGCNN, and each node is encoded with its atomic properties such as electronegativity, covalent radius, etc. The connectivity is determined based on the distance between each atom. The graph Laplacian *L* is then obtained by performing graph Fourier transformation on the crystal graph. The transformation process can be decomposed by a complete set of eigenvectors U and the eigenvalue matrix LAMBDA(λ), which represents the topology of the graph. The spectral filter $g_\theta(\lambda)$ generates a customized convolution kernel on the graph in space which can be formulated by

$g_\theta(\lambda)$=(sum from k=0 to k=k-1)$\sum\theta_k\lambda^k$

Where *k* can be interpreted as the number of convolutions in space.

However, this graph convolution cannot exploit the topological property of the graph. It is possible that the disconnected nodes have larger correlation than those connected nodes. To achieve the capability to learn from the geometric structure, we adapted a new spectral filter from Li et.al[1] which takes a non Euclean metric so that it learns a residual graph apart from the intrinsic crystal graph which makes a correction on the original graph Laplacian *L*:

L(with hat) = L + αL_{res}

The overall intrinsic plus residue graph is trained with the spectral graph convoluted neural network to predict chemical properties of crystal and results are compared with various existing models.

[1] Li, R., Wang, S., Zhu, F. and Huang, J., 2018, April. Adaptive graph convolutional neural networks. In Thirty-second AAAI conference on artificial intelligence.

599718 Interpretable Model for Molecular Data Fusion

Abstract:

In this work, we introduce a graph-theoretical framework for fusing thermochemical data across different levels of quantum theory. We show that it can be used to predict *different thermochemical quantities at a higher level of theory using a quantity at a lower level of theory.* The generalizability of the model is investigated, and rigorous statistical tests are used to guarantee bounds in model predictions. Two important aspects of machine learnt models are addressed: domain knowledge and interpretability. We show how our model draws from chemical knowledge and thus provides an interpretable mapping across levels of theory and quantities of interest. These us to draw physical insight from the learning process. The approach is illustrated with multiple mapping tasks, and levels of quantum theory for a dataset of ~12k molecules. Chemical accuracy (1 kcal/mol) is approached for one task and is surpassed for all the others. Our approach provides a blueprint of how to merge disparate and incomplete literature datasets, built using different levels of theory, to create a more comprehensive thermochemical database for applications such as Big Data analysis.

601143 Leveraging Structure and Property Information for Building Maps of Materials

Abstract:

The application of machine learning techniques to problems in chemistry and materials science has become increasingly common over the past few years, with many studies focusing on property prediction or structural comparisons based on mathematical representations of materials. This has been made possible by the emergence of large, diverse databases of materials and molecules that allow for statistical analyses and high-throughput searches of candidate materials for a particular application. In many cases, such analyses are facilitated by the construction of a "map" that helps visualize the similarities and differences between the structures and properties of materials. Such maps are often based on a reduced structural representation obtained through the use of, for example, principal component analysis (PCA). However, maps constructed in this way do not necessarily reflect the property correlations between different structures. Similarly, the representation that yields the best property predictions may not be the best for highlighting structural relationships between different materials. In this work, we propose a kernel-based extension to Principal Covariates Regression (PCovR) [1], a method that combines PCA and linear regression, to create maps of a variety of materials, including molecular crystals and all-silica zeolites, that strike a balance between predictive ability and emphasizing structural differences. The kernelized extension, which we call Kernel Principal Covariates Regression (KPCovR), allows us to take advantage of nonlinear relationships in the data to build representations of materials that can offer improved regression performance over linear PCovR while preserving or enhancing the apparent structural diversity among the materials studied. Finally, we introduce a feature selection procedure based on PCovR aimed at identifying the features that best represent the structural diversity and predictive ability simultaneously, which makes it possible to achieve more accurate property predictions with a smaller number of features compared to other simple feature selection methods.

[1] S. de Jong, H. A. L. Kiers, *Chemomet. Intell. Lab. Sys. 14*, 155-164 (1992)

597611 Determining the Structure of Organic Rosette Nanotubes Using Multi-Scale Molecular Modeling: Ring Stack Vs Helical Tube

Abstract:

Rosette Nanotubes (RNTs) are tubular soft materials self-assembled from guanine-cytosine (G∧C) hybrid molecules, which can be covalently functionalized for use in various applications. They have been shown to be effective assemblies in regenerative medicine, drug display and delivery, and catalysis.

Previous spectroscopic studies suggest they have a structure which is formed by hexameric rings, maintained by self-complementary hydrogen bonds, that are further stacked and supported by π-π interactions, which form the ring-stacked RNTs. While this mode of association maximizes the hydrogen bonding interactions and results in efficient π-π stacking, it is also possible to envision that the G∧C modules assume a helical organization defining a tubular core (Fig. 1). Having knowledge of the structure of the RNT is important in the design and synthesis of novel motifs for various applications.

Imaging the nanotube using scanning tunneling microscopy (STM) proved inconclusive in differentiating the three configurations (Fig. 1). Furthermore, comparing the chemical shifts of the nitrogen atoms of the G∧C monomers of both the stacked-tube (ST) and helical tube (HT) configurations using quantum chemical calculations showed no significant difference between them.

We further investigated this possibility by using the lysine-functionalized RNT (K1-RNT) and applying multi-scale molecular modeling methods, which include Monte-Carlo conformational search, molecular dynamics, and the statistical mechanical theory of molecular liquids (3D-RISM theory). We considered three structures of the K1-RNT: ST, left-handed HT (LHT), and right-handed HT (RHT) (Fig. 2).

Our results suggest that the formation of ST, LHT, and RHT K1-RNTs in water are favorable and are enthalpically driven. Moreover, 3D-RISM analysis suggests that the RHT conformation is more probable than the ST and LHT RNTs and this is due to a more favorable solute-solvent interaction energy (Fig. 3). Using the 3D distribution of solvent sites around the RNTs (from 3D-RISM calculations), we were also able to determine the solvation structure and estimate the free energy of binding of each water molecule in the RNT channel (Fig. 4). The results also corroborate the higher probability of the RHT conformation occurring. Studies are being done to verify these findings.

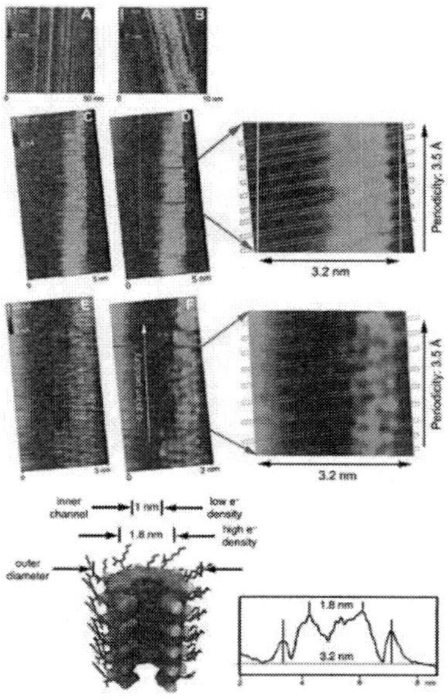

Fig. 1. STM images of K1-RNT

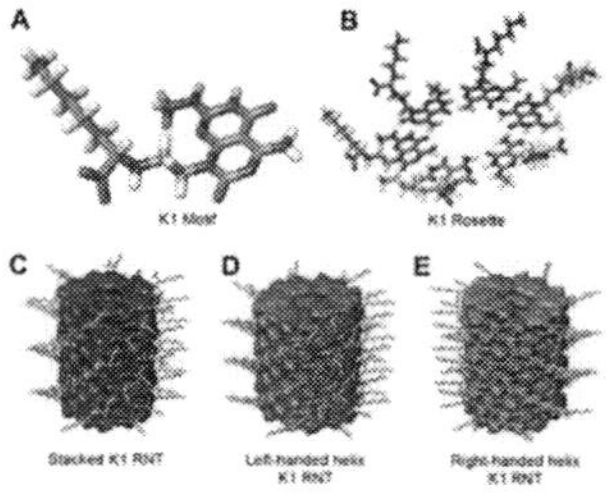

Fig. 2. Molecular models of G∧C module K1 (K1 Motif) (A) self-assembled rosette (K1 Rosette) (B) and corresponding K1-RNT in a stacked (ST) (C), left-handed helical (LHT) (D), and right-handed helical (RHT) (E) assemblies.

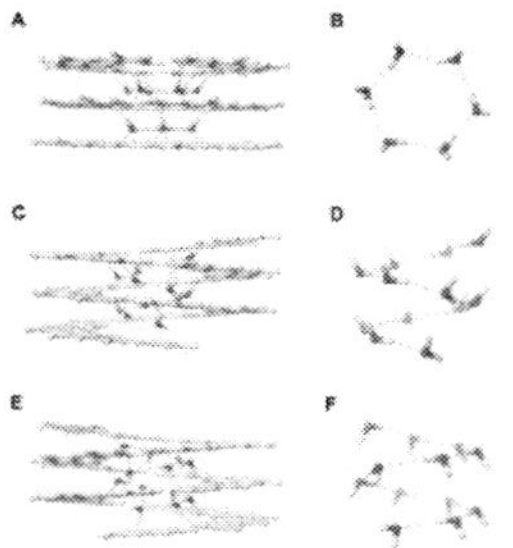

Fig. 4. Structural water inside the ST(A), LHT(C), and RHT(E) K1-RNTs and water structure inside the corresponding channels (B, D, F).

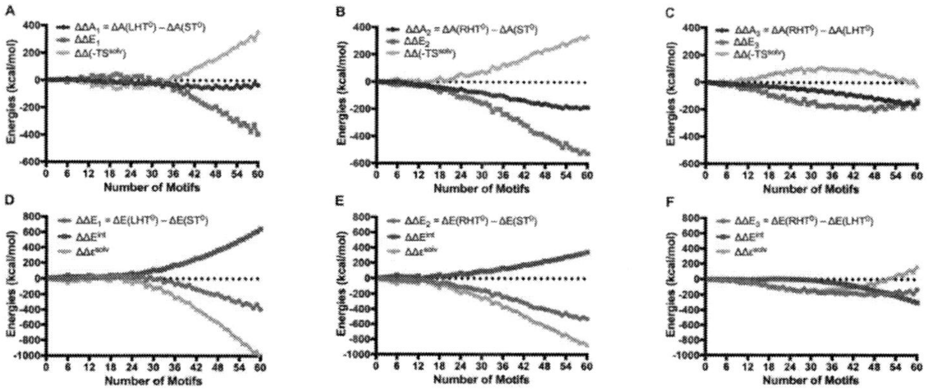

Fig. 3. Difference in association free energy $\Delta\Delta A$ and its components ($\Delta\Delta A = \Delta\Delta E + \Delta\Delta(-TS^{solv})$) of the LHT and ST (A), RHT and ST (B), and RHT and LHT (C). $\Delta\Delta E$ and its components ($\Delta\Delta E = \Delta\Delta E^{int} + \Delta\Delta\varepsilon^{solv}$) between LHT and ST (D), RHT and ST (E), and RHT and LHT (F).

593724 Understanding Unexpected Hydrogen-Bonding Patterns in Proteins with Wavefunction Theory

Abstract:

Hydrogen bonds (HBs) play a key role in influencing protein-ligand binding, controlling catalytic outcomes, and determining macromolecule structures, which in turn determine their biochemical functions. Cooperative hydrogen bonding, where the donor concurrently acts as an acceptor, is the strongest HB interaction and has been observed among certain amino-acid sidechains in biomolecules. The transient presence or absence of this particular interaction can influence key outcomes in enzymes, such as the reaction rate or selectivity. In order to understand this, it is important to develop an accurate model of how often single hydrogen bond donors/acceptors are formed versus those that involve multiple hydrogen bonds between amino-acid sidechains. We use accurate coupled cluster theory models extrapolated to the complete basis set limit to build a potential energy surface (PES) of these competing types of hydrogen bonds. We use small model systems of N-H···O or O-H···O HBs and their doubly bonded forms to ensure that we can extensively study the relevant PES. We then compare the observations on these systems (e.g., acetamide-methanol) to the representative protein amino acids. We extract representative amino acids with oxygen-atom-containing sidechains (i.e., tyrosine, serine, or threonine) hydrogen-bonded to the nitrogen/oxygen-atom-containing sidechains of asparagine or glutamine from high-resolution (1.5 Å or better) X-ray crystal structures in the Protein Data Bank. Analysis of these high-resolution protein structures reveals that the maxima of O-H···O and N-H···O HB distance and angle histograms coincide with the interaction energy minima obtained from our calculations of model systems. This comparison suggests that we have achieved quantitative prediction of the relative favorability of the hydrogen bonding motifs, despite the approximate nature of the model systems. Comparison of representative systems revealed distinct behavior for interactions between aromatic tyrosine and sidechains versus those with serine or threonine. This study provides essential insight into understanding how noncovalent interactions in enzyme active sites guide catalytic outcomes.

610423 Solubility Prediction of Organic Molecules Using Atomistic Simulations

Abstract:

More than 90% of small molecule drugs are delivered in crystalline form. The "bioperformance" and manufacturing of these crystalline molecules is a strong function of their polymorphic form, shape, and size. To engineer crystals into their most bioavailable form and with least processing difficulty, we must be able to model their form, shape, and size as a function of practical design parameters. Solubility is a crucial physical property for bioavailability, for growth rates, and nucleation rates. We have developed a new chemical potential route to solubilities, based on absolute chemical potentials in the solid and solution phases. Our gas phase chemical potential calculations begin from a novel centroid reference system, which we convert to the real molecule and then to a solvated molecule via thermodynamic perturbation and/or thermodynamic integration. The solid chemical potential is obtained by transforming the Einstein crystal to a molecular crystal (Frenkel-Ladd calculations). We illustrate the calculations by predicting the solubility of succinic acid (with GAFF force field and TIP3P water). We use direct coexistence simulations to make a self-consistent test of our chemical potential calculations.

601583 Mechanistic Insights into the Enantiotropic Layer Slip Polymorphic Transition in Paracetamol

Abstract:

Reversible temperature mediated solid phase changes, otherwise known as Enantiotropic transformations, often occur in many molecular crystals. These transformations occur as a result of free energy stabilization through entropic contributions at finite temperatures. These phase transitions can often have significant implications on the quality, performance, and characteristics of crystalline solids. As such, understanding and predicting these transformations is of great importance.

In this study, we utilize molecular simulations to elucidate the mechanisms behind the enantiotropic layer slip phase transformation between the Paracetamol orthorhombic and monoclinic versions of Form III (Form III-m and III-o), as reported by Reiss et al.[1] Using standard molecular dynamics (MD) in addition to Crystal Adiabatic Free Energy Dynamics (Crystal-AFED)[2], an enhanced sampling approach, we demonstrate that the transformation from the monoclinic Form III-m to orthorhombic Form III-o is driven by localized and dynamic disorder within the structure rather than a perfect crystal to crystal transformation. These results suggest that the orthorhombic Form III structure does not exist as a perfect orthorhombic crystal with perfectly aligned layers, but rather as a collective ensemble average of various layer slipped structures. In addition, we demonstrate the utility of standard MD and Crystal-AFED to successfully map out the free energy profile of this enantiotropic transformation as a function of temperature, and also for incorporating dynamic structural disorder in crystal structure prediction.

RH, AM, and AS are employees of AbbVie and may own AbbVie stock. MT, LVM and EC are employees of New York University and have no additional conflicts of interest to disclose. AbbVie and NYU sponsored and funded the study; contributed to the design; participated in the collection, analysis, and interpretation of data, and in writing, reviewing, and approval of the final publication.

1. Reiss, C. A.; Mechelen, J. B. v.; Goubitz, K.; Peschar, R., Reassessment of paracetamol orthorhombic Form III and determination of a novel low-temperature monoclinic Form III-m from powder diffraction data. *Acta Crystallographica Section C* **2018,** *74* (3), 392-399.
2. Yu, T.-Q.; Tuckerman, M. E., Temperature-Accelerated Method for Exploring Polymorphism in Molecular Crystals Based on Free Energy. *Physical Review Letters* **2011,** *107* (1), 015701.

604061 Molecular Dynamics Simulation Based Approach to Study Cryoconcentration of Solutes during Freezing and Its Impact on Protein Stability

Abstract:

Purpose:

Bulk protein therapeutics are generally frozen for long term storage to prevent product degradation and facilitate product transportation. The freezing process of bulk protein solutions leads to cryoconcentration of solutes, which may impact protein stability. However, the underlying mechanism and molecular details of proteins unfolding and aggregating at the cryoconcentration microenvironment are poorly understood. Molecular dynamics (MD) simulations have been developed to investigate how cryoconcentration of different solutes affects protein stability during the bulk freezing process. The simulations results from this current work have been validated with previous experimental data on solutes cryoconcentration acquired by our group.

Methods:

The initial conditions applied into the MD simulations were selected based on previous cryoconcentration experiments in which 1 L bottles containing 600 mL of a protein solution of 0.5 mg/mL lactate dehydrogenase (LDH) in 10 mM phosphate buffer at pH 7.0 were frozen at -30°C. Therefore, LDH and sodium phosphate salts' molecules were added into the simulation box. The MD simulation trajectories and their analyses were carried out using GROMACS, VMD and Pymol software. We applied both CHARMM and MARTINI force-field for all-atom as well as coarse-grained simulations, respectively. Moreover, Steered MD simulation has been applied to investigate potential mean force of protein stability related association and dissociation process. The initial coordinates for LDH were taken from the protein data bank entries (pdb: 5NQB). The protein was initially modeled using the psfgen tool in VMD, and then further solvated in a simulation box with TIP4P ice water model and charge neutralized with NaPO4. The box volume was optimized in the NPT ensemble by initially applying 500 cycles of a conjugate-gradient minimization scheme followed by a short 40-ps MD run in NPT assemble. The temperature was controlled using the Langevin thermostat, and the pressure was controlled by the Nose-Hoover barostat. The simulations were carried out using periodic boundary conditions. These briefly equilibrated systems of LDH protein were further subjected to MD simulation sets in the NVT ensemble. Computations were performed on HPC, a supercomputer cluster at the University of Connecticut.

Results:

The cryoconcentration effect was evaluated applying a model protein solvated in a simulation box of aqueous solvent in which different ions and stabilizers were added for case studies. Analyses of the resulted trajectories reveal the effect of temperature, ionic strength, and solutes concentration on the protein unfolding state, biomolecular interactions, and protein aggregation. The final concentration validation show a good match of simulation results and the previous experiment data.

Conclusions:

In this present work, we have identified the relatively more conformation variations of LDH at higher temperature which leads to unstable LDH condition, whereas, the stability has been reinforced with ice formation at lower temperature. In addition, the most flexible regions on protein corresponding with the most impact residues from ice-water interface were discovered successfully. To compare two types of ice formation cases, both-side ice formation revealed relatively more stable LDH protein conformation with less solvent accessible surface area and more hydrogen bonds. LDH protein aggregation has been identified at CG-MD at higher temperature and we discovered the possible residues aggregated based on statistics probability repeats.

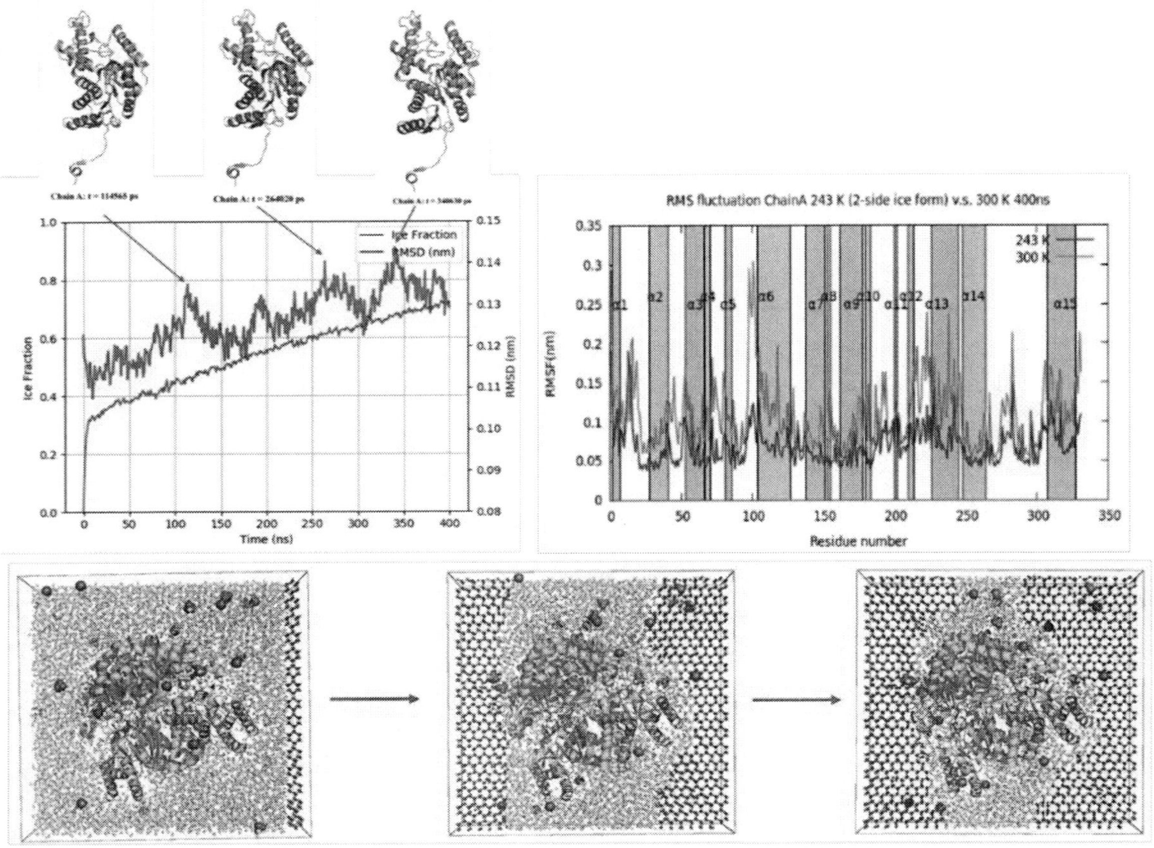

608180 Optimizing the Solvent for Extreme-Strength Two-Dimensional Polymers

Abstract:

Two-dimensional polymers (or COFs) have shown promise as candidates for next-generation extreme strength materials as they combine the advantages of traditional Kevlar and more recent graphene-like materials. In order to manufacture these materials using high-throughput solution-type processing, one must dissolve large quantities of material in a suitable solvent. In this study, we utilize quantum mechanical calculations at small length scales as well as molecular dynamics simulations at intermediate length scales to understand the nature of interactions between the 2-D polymer and different solvents. Adsorption configurations, binding energies, and charge transfer between the polymer and solvent molecules will be used to evaluate the strength of interactions and eventually, to predict optimal solvents for this material.

610845 Understanding the Specific Conductance of NaCl Solutions in Water at Elevated Temperatures and Pressures

Abstract:

Supercritical water desalination is one of the alternative processes to obtain clean water and to retrieve valuable materials. The performance of the desalination process is usually evaluated *in situ* by measuring the specific conductance (electrical conductivity). However, the specific conductance of aqueous electrolytes at elevated temperatures and pressures are known to show a complex dependence on the process conditions, which makes it challenging to evaluate the desalination process. In order to understand how the specific conductance evolves as the process condition changes, we perform the molecular dynamics (MD) simulations of aqueous sodium chloride (NaCl) solutions at elevated temperatures and pressures (298 – 674 K at 200 bar). By applying the Green-Kubo linear response theory in conjunction with the spatial decomposition analysis based on the Voronoi network, we quantitatively examine how velocity autocorrelation and cross-correlation contributes to the specific conductance of the system. The contribution of the cross-correlation term between the oppositely charged ions, which arise from the formation of ion pairs, becomes significant above 548 K, where the experimental conductance data becomes its maximum. Together with the conductance analysis, we develop a cut-off free network scientific algorithm based on the Voronoi tessellation to classify ion pairs into solvent-separated, solvent-shared, and contact ion pairs. The classification result yields an accurate calculation of the ion association constant, which can also be used for the spatial decomposition of the specific conductance. Overall results suggest that the contact ion pair (CIP) formation is mainly responsible for the onset of the decrease in the specific conductance.

608762 Near-Infrared-Absorbing Open-Shell Donor-π-Acceptor Metal-Free Conjugated Dyes for Dye-Sensitized Solar Cells

Abstract:

Dye-sensitized solar cells (DSCs) have drawn a significant interest due to their low production cost, design flexibility, and the easy tunability of the sensitizer. However, the power conversion efficiency (PCE) of the metal-free organic dyes is limited due to their inability to absorb light in the near-infrared (NIR) region, leaving a large amount of energy unused. Herein, we have designed new DSC dyes with open-shell character, which significantly red-shifts the absorption spectra from its counterpart closed-shell structure. A small radical character ($y < 0.10$) is found to be beneficial in red-shifting the absorption maxima into the NIR region and broadening up to 2500 nm. Also, the open-shell dyes significantly reduce the singlet-triplet energy gaps (ΔE_{ST}), increase the total amount of charge transfer to the semiconductor surface, reduce the exciton binding energy, and significantly increase the excited-state lifetimes than the closed-shell dyes. However, the closed-shell dyes have large injection efficiency with increased intramolecular charge transfer (ICT) character. Furthermore, the open-shell dyes found more closely adsorbed into TiO_2surface than the closed-shell dyes.

612761 Iterative Peptide Biomaterials Discovery with Maximum Entropy Methods and Deep Learning

Abstract:

The use of artificial intelligence for molecular design has been rapidly advancing. New generative models propose novel molecular structures to test and deep learning can predict their molecular properties with high accuracy. The goal for these algorithms is optimizing a property by training against a large database of previously measured molecules. A major limitation is that these methods are not designed to be used concurrently with experiments and improving their accuracy as experimental data is gathered is non-trivial. Our research is focused on iterative materials design where experiments are done concurrently instead of post-hoc in molecular design. This is accomplished by using maximum entropy biasing methods that can update predictive physics-based models with new experimental results. This improves predictive accuracy and ensures interpretable models. Another challenge is the scarcity of data in materials design which we're addressing with meta-learning. Meta-learning translates experience from past related systems to new ones, minimizing the number of molecules necessary to train models. I will present results of these methods applied to peptides where we have studied a variety of tasks including antimicrobial, antifouling, and solubility predictions. Peptides are a model system because there is no ambiguity in encoding and generating them because they are linear biopolymers made of 20 amino acid monomers.

601160 Discovering Molecules with Selective Membrane Partitioning Using Active Learning and Alchemical Transformation

Abstract:

In this work we integrate coarse-grained molecular dynamics simulation, alchemical free energy calculations, deep representational learning, and Bayesian optimization to discover to discover small organic molecules that selectively permeate into cardiolipin membranes. The phospholipid cardiolipin is exclusively found within the inner mitochondrial membrane in eukaryotic cells and is intimately involved in mediating metabolic and regulatory cascades. Deviations in cardiolipin composition are indicative of a number of pathologies, such as Tangier disease. Identifying molecules that selectively penetrate into membranes composed of cardiolipin to probe the pool of cardiolipin *in vivo* can serve as a diagnostic tool to identify pathologies in mitochondrial membranes. To effectively screen the space of small organic molecules less than ~450 Da, which are likely to passively permeate without the assistance of membrane transporters, we computationally search the molecular design space using molecular simulation and an active learning-directed molecular design platform. Candidate molecules are evaluated using alchemical transformations to compute the free energy of inserting into lipid bilayers of cardiolipin relative to those of POPG, a common phospholipid found throughout the eukaryotic cells. Deep representational learning is used to embed the discrete molecular design space into a smooth, low-dimensional latent space within which we conduct an active learning search for highly selective molecules using Gaussian process regression and Bayesian optimization. Using this platform, we identify a small number of novel candidate small molecules predicted to have superior selectivity for cardiolipin membranes relative to current state-of-the-art diagnostic drugs.

608973 Integrating Coarse-Grained Modeling, Machine Learning, and Optimization for Biopolymer Design

Abstract:

The chemical design of polymers with target structural and/or functional properties represents a grand challenge in materials science. While data-driven design approaches are promising, success with polymers has been limited, largely due to limitations in data availability. In this talk, I will discuss a promising workflow that combines coarse-grained modeling, machine learning, and model optimization to achieve the targeted sequence design of biopolymers. This workflow will be explained for the test case of controlling polymer folding/size based on polymer sequence/chemistry, and early results towards enzyme encapsulation will be presented. Important facets of machine learning for polymer applications, such as featurization strategies and data requirements will also be discussed.

609627 Machine Learning for the Prediction of Glass Transition Temperature of Biodegradable Polymers

Abstract:

The glass transition temperature (T_g) is the temperature at which the amorphous polymer changes its state from glassy to rubbery. T_g depends on the structural features of the polymers. These structural features include cross-linking in the polymer structure, polymer chain stiffness, molecular weight, number of carbon atoms etc. In general, chemical groups and intermolecular forces that are present in the polymer structures are the main factors that influence the T_g of polymers. This specific information about the structural features can be numerically encoded as molecular descriptors, a unique value associated with the chemical constitution for correlation of chemical structure with various physical properties. The purpose of the quantitative structure – property relationship (QSPR) studies is to find the relationship between the structure and the property of interest using these molecular descriptors.

In this study, we are using machine learning models such as multiple linear regression (MLR), support vector machine-based regression, random forest regression, K- nearest neighbors regression (KNN) and artificial neural networks (ANN) to construct a QSPR model. The QSPR model will be applied to the prediction of T_g for 59 biodegradable polymers having low T_g (from -35°C to 225 °C), 85 synthetic nondegradable polymers with moderate T_g (from 250 °C to 399 °C), and 77 aromatic polymers with high T_g (from 500 °C to 800 °C). The chemical structures of all the polymers were drawn in ChemDraw 2D and exported to ChemDraw 3D. The 3D structures were then optimized by applying energy minimization using molecular mechanics (MM2) until the rms gradient value became smaller than 0.1 kcal/mol Å. The optimized structures were further optimized using MOPAC method until rms value was reduced to smaller than 0.0001 kcal/mol Å. 1825 molecular descriptors were generated from the optimized chemical structures using the algorithms in the Mordred web graphical user interface (GUI) of python. Out of 1825, six descriptors were selected for machine learning models by means of stepwise regression followed by least absolute shrinkage and selection operator (Lasso) regularization methods using MATLAB. Using these six features the proposed machine learning models are being trained on 80 percent of the dataset created using Mordred GUI and tested on remaining 20 percent of the dataset for the prediction of T_g. We are also investigating the classification techniques such as decision trees, random forests, and support vector machine classifier (SVC) to classify polymers into different classes based on the relation between the selected molecular descriptors and T_g.

Keywords: Glass transition temperature, Biodegradable polymers, QSPR, Machine learning.

611100 Integration of Metabolic Network and Data-Driven Modeling to Predict Context-dependent Microbial Interactions

Abstract:

Interspecies interactions provide microbes with an important means to promote their fitness so that they can survive in a broader range of environmental conditions as a community, while not individually. As a prime survival strategy, therefore, microbes dynamically modulate their interactions and even change their partners subject to environmental variations. A fundamental understanding of this context-dependent interaction is key for accurately predicting community dynamics and function. Dynamic Flux balance analysis (dFBA) provides mechanistic predictions of how metabolic interactions occur among species through cross-feeding and competition and how the shifts in interactions can be context-dependent across varying environments. Despite this usefulness, mechanistic models alone cannot *quantify* net interactions among species and their temporal variations, particularly when the system is characterized by *mixed* interactions. To compensate for this limitation, we propose the integration of data-driven modeling with dFBA. We used Sparse Identification of Nonlinear Dynamics (SINDy) to infer microbial interactions with each other and with the environment. Unlike conventional network inference tools that assume constant interaction coefficients, SINDy allows for the identification of interaction coefficients that are dependent on species abundances and environmental variables. In the case study of a co-culture of auxotrophic *Escherichia coli* strains, integration of SINDy with dFBA provided not only accurate predictions of the temporal shifts in interactions as the limiting nutrients change in time, but also mechanistic interpretations of how these shifts occur in metabolic reaction levels in individual species. Beyond simple co-cultures, the proposed method can be extended to model more complex communities.

605677 Molecular Design of Underwater Adhesives Using DOPA and Amyloid-Forming Peptide Conjugates

Abstract:

Adhesives capable of sticking to multiple types of surfaces in underwater and high moisture conditions are required for a large array of applications such as marine coatings, sealants, medical devices, and wet living tissue repair. Most typical industrial adhesives, however, are either developed primarily for dry applications and perform poorly in wet environments or are highly specific to particular surface types and not suitable for broad use. A viable and cost-effective strategy to address this issue is to develop generic adhesives that can be used for various surfaces in water or high moisture conditions.

One approach researchers have taken to develop new adhesives for these applications is to draw inspiration from the natural glues produced by aquatic organisms capable of strong, moisture-resistant adhesion to many different surfaces. Analysis of these natural marine glues has shown that these organisms' abilities to adhere to a multitude of surfaces involve L-3,4-dihydroxyphenylalanine (DOPA) or functional amyloid nanostructures.

The objective of this project is to computationally design bio-inspired underwater adhesives by analyzing and combining the adhesive capabilities of DOPA and amyloid-forming peptides in wet conditions. Explicit solvent atomistic simulations can give insight into the interactions underlying surface adhesion in the presence of water and implicit coarse-grained simulations can show the formation of peptide structures, which can then be used to guide the design of DOPA-amyloid conjugates for underwater adhesion.

In this study, we begin the process of designing a DOPA-amyloid conjugate capable of generic underwater adhesion by examining the interactions between DOPA and a model surface and then by testing various DOPA-amyloid conjugate designs for their amyloid forming propensity. We first investigated the basic interactions between DOPA at a model underwater silica surface and found that DOPA readily adsorbed onto the model surface. In our DOPA-amyloid conjugate designs, the amyloid-forming peptide KLVFFAE was chosen for the initial design as it is known to self-assemble into functional amyloid nanostructures. Simulations of the conjugated DOPA-amyloid chains showed that the DOPA motifs did not interact with the rest of the chain, which leaves them free to bind to the surface. After several designs were tested for their amyloid-forming propensity, it was found that the sequence KLVFFAE-G-ddGddGdd (where d represents DOPA) was the most favorable of the currently tested designs. Future work will focus on furthering understanding of the behavior of DOPA-amyloid chains at model surfaces and how changes in the backbone of the DOPA-containing chain and in the sequence of the amyloid-forming peptide affect adhesion on the model surfaces.

608696 Intelligent Design of Substituted Ferrite Materials for Magnetically Modulated Energy Delivery Applications: Density Functional Theory and Machine Learning Study

Abstract:

We use computational methods to drive the design of substituted ferrites. Substituted ferrites are magnetic materials based off of the bulk structure of magnetite, i.e., Fe_3O_4 (**Figure 1a**). In this structure, two of the Fe ions are octahedrally coordinated with formal charges of +2 or +3, and one is tetrahedrally coordinated with a formal charge of +3 (**Figure 1b**). Substituted ferrites have the formula $M_xFe_{(3-x)}O_4$, where M is the substituent and can be Mn, Co, Ni, Cu, or Zn. Capitalizing on their strong and tunable magnetic properties (e.g., magnetic saturation and magnetic anisotropy), substituted ferrites can be used for energy delivery when placed in a magnetic field. For biomedical applications, a goal is to optimize the magnetic saturation and anisotropy so that energy can be delivered (e.g., to cancer cells) effectively and safely. Recent literature suggests that the magnetic properties of substituted ferrites can be tuned via their compositions. In this work, we use a combination of Density Functional Theory (DFT) and machine learning (ML) to learn the relationship between composition and magnetic performance of substituted ferrites. Magnetic saturations and anisotropies are calculated for ~1,000 substituted ferrite bulk structures (with x ranging from 0.0625 to 1) using DFT. Substitutions are made at the octahedrally coordinated Fe^{2+} sites, which are highlighted in gray in **Figure 1b**. DFT values are input to machine-learning (ML) to correlate the magnetic properties to composition. Specifically, an XGBoost algorithm is used to correlate the magnetic performance to tabulated properties of the substituents, such as electronic properties, nuclear properties, atomic sizes, and redox energies. All of these "features" of the substituents are taken from established databases and do not require further calculations. **Figure 1c** shows a parity plot comparing the magnetic saturation as computed with DFT and predicted with our ML model, and **Figure 1d** is a radar chart illustrating features selected with the Recursive Feature Elimination with Cross-Validation (RFECV) algorithm that have the strongest influence on the magnetic saturation. The developed ML model enables the screening of vast chemical spaces with surrogate ML models, which can be used to identify design rules and guide materials selection for substituted ferrites.

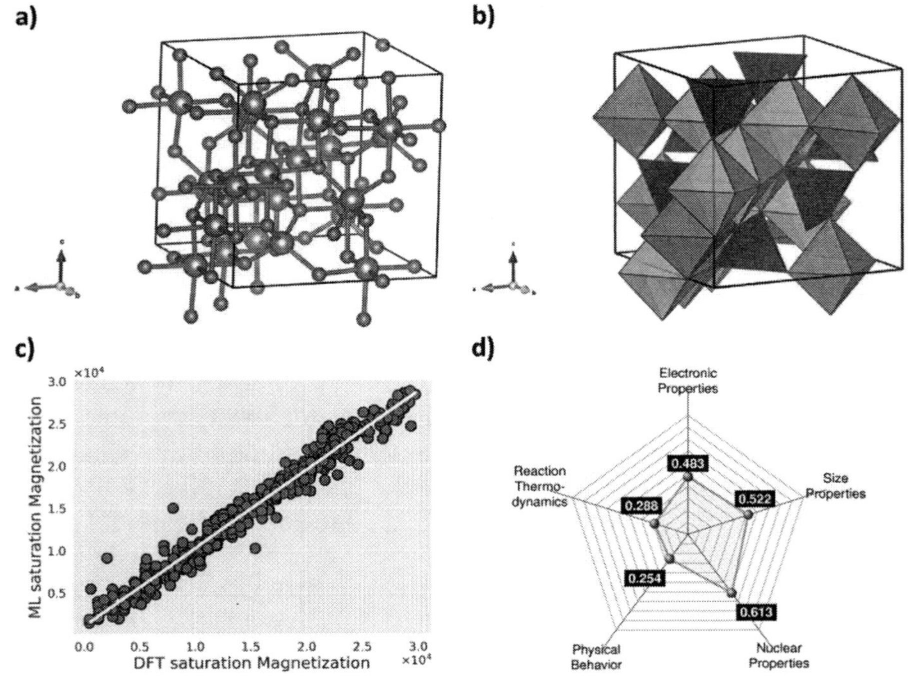

Figure 1. (a) Ball-and-stick style rendering (Fe and O atoms are colored in tan a red), b) Polyhedral style rendering—purple, orange and gray polyhedral represent tetrahedral (Fe^{2+}), octahedral (Fe^{3+}), and octahedral (Fe^{2+}) sites respectively, c) parity plot comparing the saturation magnetization computed from DFT and ML model, and d) radar plot with the feature selection analysis.

598112 Computational Modeling of Therapeutic Glycoproteins

Abstract:

Glycosylation is an important posttranslational modification and plays a crucial role in protein efficacy. It is crucial to understand protein glycosylation in order to optimize recombinant protein expression. We study 3D structures of glycoproteins in atomistic detail starting from known crystal structures to examine how different glycosylation profiles can affect glycoprotein structure and function. Novel computational methods for modeling glycoprotein structures are employed to understand how changes in site-specific oligosaccharide composition and structure influence 3D glycoprotein structure, protein oligomerization, and product quality attributes such as activity/efficacy, stability, and immunogenicity. We are applying this methodology to a number of different proteins: Butyrylcholinesterase, a scavenger for organophosphorus nerve agents,; CMG2-Fc, a plant-made Fc-fusion anthrax decoy protein, and ACE2-Fc, a (plant-made) Fc-fusion protein that binds the Sars-Cov2 spike protein.

599842 Deep Learning of Low-Dimensional Latent Space Molecular Simulators

Abstract:

The long-time microscopic evolution of molecular systems is governed by the leading eigenfunctions of the transfer operator that propagates the system through time. The low-dimensional latent space defined by these eigenfunctions parameterize the slow manifold to which the system is constrained to evolve. We have developed deep neural networks of various architectures to (i) learn the leading transfer operator eigenfunctions and project molecular configurations into this latent space, (ii) propagate the molecular dynamics within the encoded latent space, and (iii) decode the latent space back to the all-atom coordinate space. We train the networks over short simulation trajectories to establish low-dimensional latent space simulators capable of performing molecular dynamics simulations at orders-of-magnitude lower cost than traditional physics-based simulation. In an application to Trp-cage protein, we train latent space simulators over 200 μs of traditional physics-based simulation trajectories costing millions of CPU-h to generate. Our trained model can generate simulations of equal length in just 2.3 s of wall time on a single CPU core, and the synthetic trajectories are in excellent accord with the structure, thermodynamics, and kinetics of the original simulations: our all-atom configurations are lie within <5 Å RMSD and our free energy landscapes and kinetic rates are in quantitative agreement. The trained latent space simulators can generate thousands of folding and unfolding events in a matter of minutes, enabling us to quantify long-time kinetic phenomena at unprecedented statistical accuracy while maintaining a continuous-time, continuous-space, all-atom representation of the molecular dynamics.

607296 Mosdef Cassandra: A Python Interface Providing Complete Integration between the Cassandra Monte Carlo Software and Mosdef Tools

Abstract:

Over the previous half-century, molecular simulations have proven an invaluable technique for investigating and predicting the behavior of matter. The Monte Carlo (MC) methods are particularly valuable for their ability to easily simulate open ensembles, i.e., including molecule insertions and deletions. This capability enables the simulation of phase equilibria, such as vapor-liquid equilibrium and adsorption in porous materials. Rapidly increasing computing power makes MC simulation, in principle, more powerful. However, generating input files and managing the overall simulation workflow still requires human intervention and suffers from human error. The next generation of simulation studies will likely include more high-throughput screening and perhaps even iterative approaches that combine statistical methods (e.g., machine learning) and molecular simulations to probe the behavior of chemical systems across large parameter spaces. These types of studies will require two key components from the next generation of MC simulation software: enhanced simulation reproducibility and improved integration with other community-based software tools.

In this work, we demonstrate our recent integration of the *Molecular Simulation Design Framework* (MoSDeF) [1] with *Cassandra* [2] via a new Python interface for *Cassandra*. MoSDeF is an effort to promote TRUE simulation workflows (Transparent, Reproducible, Usable by others and Extensible) by providing modular Python-based tools that can be used during the system creation and force field application steps of a simulation workflow. The goal of these tools is to remove the use of graphical user interfaces, in-house interconversion scripts, and other ad-hoc steps used by many researchers to create and modify input files for molecular simulations. The Python-based Cassandra interface offers three main advantages: improved reproducibility by containing the entire simulation workflow within a single script, a tremendously simplified process of performing an MC simulation, and easier integration with other software tools.

We describe the philosophy and design of the *MoSDeF Cassandra* interface and demonstrate how the software simplifies the act of performing an MC simulation. Starting from the "traditional" *Cassandra* simulation workflow, we show how *MoSDeF Cassandra* removes ad-hoc modifications and the need for graphical user interfaces. Instead, the entire simulation workflow (system creation, force field assignment, simulation, and analysis) is contained in a single Python script. This script is easy to version-control and share. Reproducing a simulation is as trivial as downloading the script and installing the appropriate software. Modifying the simulation (e.g., different conditions, molecules, etc.) is often nearly as trivial. We then demonstrate extensions to more complex workflows, such as calculating adsorption isotherms in porous media. Finally, we show how *MoSDeF Cassandra* integration with *Signac* [3] enables simplified job management and submission for thousands or more simulations. With this power, we believe MoSDeF Cassandra is well-positioned to enable the next generation of Monte Carlo simulations.

[1] M.W. Thompson, J.B. Gilmer, R.A. Matsumoto, C.D. Quach, P. Shamaprasad, A.H. Yang, C.R. Iacovella, C. McCabe, P.T. Cummings. Towards molecular simulations that are transparent, reproducible, usable by others, and extensible. Mol. Sim. *In press,* (2020).

[2] J.K. Shah, E. Marin-Rimoldi, R.G. Mullen, B.P. Keene, S. Khan, A.S. Paluch, N. Rai, L.L. Romanelo, T.W. Rosch, B. Yoo, E.J. Maginn, Cassandra: An open source Monte Carlo package for molecular simulation. J. Comput. Chem. 38 (2017) 1727-1739.

[3] V. Ramasubramani, C.S. Ardorf, P.M. Dodd, B.D. Dice, S.C. Glotzer, Simple data and workflow management with the signac framework. Comput. Mater. Sci. 146 (2018) 220-229.

609734 Automatic Generation of Ionic Liquid Libraries for High Throughput Screening

Abstract:

Ionic liquids are a class of solvents comprised entirely of bulky cations and anions. The asymmetry in the ionic structures often prevents organized packing, which results in these solvents existing as liquids at or below room temperatures. Manipulation of the cation, anion, and the substituents on the ions is the primary strategy aimed at designing a large number of ionic liquids possessing desired physicochemical, electrical, or biological properties. Probably one of the most studied substituents is the alkyl chain, especially attached to the cation. For the most part, the focus in the literature has been predominantly on the *n*-alkyl group; the isomerism embedded in the alkyl chain has been examined to a significantly lesser extent. As the number of isomers grows dramatically with the carbon chain length, there is a great potential for exploiting the isomerism to expand the number of ionic liquids.

In this presentation, we will describe an algorithm for enumerating the structural isomers of alkyl groups up to hexadecyl and the corresponding SMILES strings. We will then elaborate on our protocol for eliminating SMILES strings producing identical isomers, so that the computational effort is concentrated only on unique isomers. Next, we will describe methods for attaching the isomers to various functional groups and generating three-dimensional starting structures from SMILES strings by taking the 1-alkyl-3-methylimidazolium family of cations as an example. Results of the frontier energy orbitals obtained from electronic structure calculations will be presented and analyzed for patterns based on the respective Tanimoto similarity index.

609374 Gixstapose: Interactive and Reproducible Scattering Analysis of Simulation Snapshots

Abstract:

We develop a tool for solving common problems in structural analysis at nanometer length scales. Molecular simulations typically employ periodic boundary conditions to represent conditions in bulk materials, but this places constraints on observable structural features. Analyzing material structure in the frequency domain quantifies the degree to which repeating structural features exist, but can be difficult to implement, interpret, and reproduce. Furthermore, the generation of simulated diffraction patterns from molecular simulation plays a crucial role in validating simulation results against experiments that rely on diffraction patterns to quantify structure.

In this work we develop GIXStapose, a new interactive analysis software for studying crystalline and amorphous materials structures. It enables grazing incidence X-ray scattering (GIXS) patterns to be visualized while interactively rotating chemical structures, especially periodic simulation volumes generated from molecular simulations. This functionality is useful for interactively identifying real-space chemical features that correspond to bright diffraction peaks and the rotation matrices that generate them. As such, this tool has potential to aid in the reproducible generation of publication quality figures which connect both GIXS and structural data, and it has pedagogical potential for students learning about crystal structures and diffraction.

GIXStapose is made possible by open-source packages, including the high-quality rendering of the Fresnel ray-tracer, the chemical file formats parsing of MBuild, and numpy's fast Fourier implementations used in interactive diffraction analysis, and by funding from the National Science Foundation (#1835593)

610963 Automated Calculation of Coarse-Grained Potentials Using the Iterative Boltzmann Inversion (IBI) Method

Abstract:

Computation at mesoscopic length scales between the atomistic and continuum for polymeric and similar soft materials requires coarse-graining (CG) techniques. These techniques enable computational access to the greater length and time scales inherent to these materials by subsuming atoms associated with monomers or other chemical moieties into “coarse-grained” sites and then describing the physics by means of a potential parametrized for the coarse-grained description of the material. The development of methods to properly parametrize coarse-grained potentials is an active area of research compromising many different approaches across the soft matter community. Systematic, "bottom-up" methods, in which the coarse-grained (CG) force-field is directly derived from all-atom (AA) simulation data, have the advantage of retaining a high degree of chemical specificity, and thus, can typically capture equilibrium structure quite well.

A widely used systematic method is Iterative Boltzmann inversion (IBI) which retains chemical structure via the radial distribution function (RDF), and hence, thermodynamic properties. In IBI, an initial guess for the CG bonded and non-bonded potential are iteratively refined by means of a correction proportional to the ratio between the atomistic (target) and coarse-grained sampling distributions. We report here on a software code which automates the development of coarse-grained potentials using IBI. The code base to support this consists of a Python driver, several Python modules with core library functions, an external C++ code for calculating distributions, and the use of LAMMPS as a molecular dynamics (MD) engine. Two major problems make automation difficult: 1) noisy distributions derived from sampling; 2) low sampling regions which introduces discontinuities in the sampling. Both of these make differentiation of the energy to calculate forces difficult and error prone. Our approach addresses these problems by the construction of library functions which combine data smoothing, fitting to Gaussian functionals (bonded potentials), and flexible extrapolation schemes to handle low sampling regions without discontinuity or data distortion. The problem of overfitting is handled in part by error handling to keep the procedure running smoothly. We also introduce a library to automate the conversion of AA to CG representations and have built in an input-output framework which aims at providing reproducibility. Several use cases will be discussed aimed at providing guidance on usage.

References

1. S. Plimpton, Fast Parallel Algorithms for Short-Range Molecular Dynamics, J Comp Phys, 117, 1-19 (1995).

606572 Designing Stable Semiconductor Nanowires Via a Generic Materials Optimization Toolkit

Abstract:

Semiconductor nanowires (NWs) exhibit distinct properties due to its unique one-dimensional (1d) periodicity [1]. They are promising materials in numerous catalytic applications [2,3,4], and have attracted extensive research focus. The fast-growing NWs fabrication ability [5] brings the new challenge of NWs engineering design. This work proposes designing highly stable NWs using our new Python-based generic materials optimization package, MatOpt [6].

Cohesive energy (CE) indicates a NW's overall stability. Consequently, the search for energetically most stable NWs can be simplified to maximizing CE. Previous work in the literature has focused on using heuristic search methods such as genetic algorithms [7] and simulated annealing [8] to find the energetically most favorable NW structure. These approaches generally find stable nanostructures but often miss some of the most stable ones. We propose here a mathematical optimization-based design framework for this combinatorial design problem. The CE structure-function relationships are cast into a rigorous Mixed Integer Linear Programming (MILP) model. The design space can then be explored systematically by solving the model via established numerical optimization algorithms. The obtained, optimally-stable NW designs can serve as model structures for further experimental and theoretical investigations.

A similar, MILP-based, bottom-up design framework has been previously demonstrated for various types of nanomaterials, including metallic nanoclusters [9,10], periodic catalytic surfaces [11,12], and bulk metallic oxides [13]. In those cases, the design framework has found unintuitive optimal solutions and revealed interesting physical trends. To simplify the implementation of our generic framework and make it accessible to non-experts in combinatorial mathematical optimization, we have codified our methodologies into the Python-based toolkit MatOpt. While we focus on using MatOpt to design 1d-periodic NWs by maximizing CE, we shall also illustrate how the methodology can be easily extended further to a broad range of NW design problem and other NW functionalities of interest.

597100 Extending the Reach of Particle-Based Simulations with Machine-Learned Models

Abstract:

Many of the most challenging problems in chemical engineering involve reactivity in complicated systems. First principles methods like density functional theory (DFT) are attractive for such studies but due to computational expense, are confined to problems of a few 100s atoms and 10s of ps. These limitations have motivated extensive work towards development of accurate, scalable, and reliable reactive force fields but these models generally rely on complicated functional forms requiring time consuming parameterization schemes. As a result, the physiochemical space over which parameters are available is limited.

In this work, we present the Chebyshev Interaction Model for Efficient Simulations (ChIMES), a new reactive force field and semi-automated fitting framework that retains most of the accuracy of DFT while decreasing computational requirements by several orders of magnitude. ChIMES models are comprised of n-body atomic interactions constructed from linear combinations of Chebyshev polynomials, and are entirely linear in fitting coefficients. Thus, model parameters can be rapidly (and iteratively) generated by force matching to short DFT trajectories. ChIMES models are particularly well suited for studies that frequently rely on costly quantum simulations for elucidation and interpretation of experiments, and to date, have been successfully applied to problems including chemical reactivity under extreme conditions and surface chemistry, which can probe time and lengths that are many orders of magnitude greater that what can be achieved with standard DFT. Moreover, by overcoming this time and length scale gap, ChIMES enables direct comparisons to experiment for the first time, in many cases. Model details will be presented as will selected applications.

This work was performed under the auspices of the U.S. Department of Energy by Lawrence Livermore National Laboratory under Contract DE-AC52-07NA27344. LLNL-ABS-808063

610207 Physically-Motivated Requirements of Machine Learning Potentials

Abstract:

Machine learning potentials (MLPs) for molecular dynamics simulations have been found to be capable of approaching *ab initio* accuracy with the computational efficiency of empirical potentials. The accuracy and performance of an MLP is highly dependent upon the choice of descriptors used to characterize the local atomic environment though, as well as on the training data that the algorithm uses to predict the energies and forces. The descriptors should respect the symmetries of the atomic environment, be differentiable with respect to the atomic coordinates, and contain sufficient information. If a Gaussian process is used for machine learning, the covariance function that quantifies the similarity or closeness between two points in descriptor space also has requirements to consider. Specifically, the choice of covariance function should impose minimal constraints on the potential to reduce the risk of systematic error. This work reports on recent progress using Gaussian process regression and a novel set of descriptors to construct accurate and efficient interatomic potentials.

607356 Many-Body Machine Learning Force Fields for Explicit Solvent Modeling

Abstract:

Modeling solvation effects in computational chemistry often requires fully explicit treatments (i.e. molecular dynamics) to accurately and reliably capture solute-solvent interactions. Unfortunately, such methods require pre-parameterized force fields or many quantum mechanics calculations that can incur a prohibitively high computational cost. Recent developments in efficiently trained symmetric gradient-domain machine learning (sGDML) force fields have shown that they can enable molecular dynamics with CCSD(T) accuracy. However, a central limitation of sGDML force fields is their lack of transferability in that they are only applicable to systems the force field has been trained on. To address this issue, we consider the applicability of using a many-body decomposition comprised of multiple sGDML models that run simultaneously. Specifically, we trained sGDML models to learn energies for monomer, dimer, trimer, and tetramer interactions based on ab initio molecular dynamics simulations of solvent clusters. We found that many-body GDML models for water, acetonitrile, and methanol have more accurate force and energy prediction when compared to standard sGDML models. Furthermore, we demonstrate that force fields with up to 3-body interactions are reasonably accurate and transferable for dynamics simulations on larger solvent clusters of arbitrary size.

598149 *Quantum Computation for Simulating Vibrational Properties of Molecules*

Abstract:

Many technologically pertinent chemical properties, including reaction rates and light absorption behavior, depend on a molecule's vibrational and dynamical behavior. But in a large subset of molecules for which vibrations are relevant, accurate simulations are intractable on classical computers. Here, we discuss the use of quantum computers to achieve exponential computational speedups, highlighting approaches designed to make optimal algorithm choices based on an arbitrary quantum computer's performance characteristics. Our numerical tests show that there is a rich relationship between the problem type and the optimal method for encoding a vibrational degree of freedom into the quantum computer. The approaches we present are relevant to running simulations on near-term quantum hardware.

613690 Spontaneous Drying of Non-Polar Deep-Cavity Cavitand Pockets in Aqueous Solution

Abstract:

There are many open questions regarding the hydration of solvent-exposed non-polar tracts and pockets in proteins. Although water is predicted to de-wet purely repulsive surfaces and evacuate crevices, the extent of de-wetting is unclear when ubiquitous van der Waals interactions are in play. The structural simplicity of synthetic supramolecular hosts imbues them with considerable potential to address this issue. To this end, we detail a combination of densimetry and molecular dynamics simulations of three cavitands, coupled with calorimetric studies of their complexes with short-chain carboxylates. Our results reveal the range of wettability possible within the ostensibly identical cavitand pockets – which differ only in the presence and/or position of the methyl groups that encircle the portal to their non-polar pockets. The results demonstrate the ability of macrocycles to template water cavitation within their binding sites and show how the orientation of methyl groups can trigger the drying of non-polar pockets in liquid water, which suggests new avenues to control guest complexation.

613692 On Methods for Computing Free Energy Differences and Barriers in Soft Materials Undergoing Ordering Transitions

Abstract:

Accurate computation of free energies plays a key role in molecular simulations aimed to elucidate the thermodynamic or kinetic behavior of complex systems involving oligomers, polymers or nanoparticles. This is particularly important in understanding disorder-to-order transitions in bulk materials and to discriminate among stable and metastable states. Ordered phases of interest include crystalline assemblies formed by nanoparticles and mesophases formed by the microphase segregation of polyphilic, multiblock molecules.

After reviewing some of the general statistical mechanical principles involved in modeling free energies and how the connection among some old and new methodological variants are often overlooked (an issue that generalized frameworks try to remedy), I will describe selected methods that our group has developed to efficiently evaluate free energy changes associated with different phases and with the transition barriers between phases. These include methods for optimizing multistage ensemble simulations, for tracing phase boundaries, and for estimating free-energy nucleation barriers for situations posing significant challenges to conventional methods. I will illustrate the usefulness of these methods through sample applications concerning the self-assembly behavior of novel particle-based crystals and oligomeric mesophases. I will conclude by pointing out some of the outstanding methodological challenges, e.g., associated with mapping and inverse-mapping the behavior of coarse and fine-grained models.

613688 ATESA: An Automated Aimless Shooting Workflow

Abstract:

Transition path sampling methods are an indispensable tool for computational study of the dynamics of rare events in molecular simulations. However, access to advanced sampling techniques is generally restricted to experts. ATESA is a new open-source software tool that automates a full transition path sampling workflow based on the aimless shooting algorithm, with a particular focus on extending accessibility to non-expert users who are familiar with the basics of molecular simulations, but not necessarily with programming or advanced sampling methods. Novel contributions to the theory are also described.

618123 Theory and application of graph neural networks for molecular modeling

Abstract:

Graph neural networks (GNNs) are models that take graphs as input. GNNs are useful for modeling molecules as graphs because they are permutation invariant, meaning they do not change output if atom order is rearranged. GNNs are revolutionizing modeling of physical systems like the convolutional neural network did for imaging. In this video, we will give an overview of the principles of GNNs. After that, we will discuss their applications in coarse-grained mapping prediction and chemical shifts prediction where in both works, GNNs show good performance. GNNs should be a drop-in replacement for previous molecular feature selection and will grow in popularity.

613691 Making Molecular Simulation As Useful As Possible in Chemical Engineering

Abstract:

Molecular modeling and simulation has the potential to be transformative in chemical engineering as a tool for both predicting properties and elucidating the nature of chemical systems difficult to investigate by experiment. As of yet, this potential has only begun to be fully realized. Historically, the lack of computational power was one of the main limiting factors to useful molecular simulation. Now, however, the most useful ways to improve our molecular modeling capabilities include helping students gain the significant expertise required to generate useful results, increasing the robustness of simulation codes, improving interoperability of molecular simulation programs, increasing validation with respect to experiment or previous simulations, and improving reproducibility across the available simulation codes. I will talk about efforts of myself and my group in these areas, including the Living Journal of Computational Molecular Science, the Open Force Field Consortium, improved formulation of statistical mechanical algorithms, and improvements to molecular simulation code robustness and capability. Finally, I will discuss how we can all work together to improve molecular simulation utility now and in the future.

613101 Recent Advances in Maximum Entropy Methods

Abstract:

Making experimental data a first-class input like temperature or pressure is the goal of maximum entropy methods. This is important where force fields are approximate, such as when employing coarse-grained models or exploring new chemistry. In this talk I will review our recent advances in maximum entropy methods including treating uncertainty, large amounts of experimental data, and distributions. Maximum entropy has been used to simulate highly accurate ab initio molecular dynamics of water, incorporate environmental fluctuations inside of a macromolecular protein complex, improve RNA force fields, and to combine enhanced sampling with minimal biasing to model peptides. Maximum entropy is seeing renewed interest beyond molecular simulation. We have shown that maximum entropy can be used to connect physics-based models like molecular simulations with AI black-box predictions. Maximum entropy is even being applied to epidemiology, where complex disease models need to match current case statistics. Maximum entropy methods have revolutionized complex models by making it possible to rapidly match data without parameter optimization.

604116 A Data-Driven Approach to Understanding and Predicting the Early Formation of the Solid-Liquid Electrolyte Interphase

Abstract:

A remaining challenge in the pursuit of rational design of novel and optimized electrolytes is understanding and ultimately predicting the reaction cascade responsible for the creation of a functional liquid-anode solid-electrolyte interfaces (SEI). We here present an overview of the challenge, in different energy storage systems and advocate that a data-driven approach coupled with a high-throughput quantum chemistry computational infrastructure[1] can address the complexity of such reaction cascade. To this end, we have constructed a chemical reaction network containing over 50,000 elementary reactions which allow for bond breaking/formation as well as oxidation/reduction reactions. The reaction energetics are obtained from our computational framework which automate geometry optimization and vibrational frequency calculations for molecules, including radical, charged, metal-coordinated, and solvated species. Our framework, using select levels of theory and a suite of on-the-fly error handlers, is able to successfully converge to a minimum energy structure and calculate the molecular thermodynamic properties in over 97% of cases. Individual calculations can then be combined to perform complex workflows in a fully automated fashion. To date, we have applied this framework to over 25,000 unique molecules relevant to the formation of solid-electrolyte interphases in Li-ion and Mg-ion batteries. To analyze possible reaction pathways, we apply a graph representation to our reaction network, leveraging an iterative reweighting strategy that solves for all prerequisite branching such that traditional pathfinding algorithms correctly capture coordinated pathways. Machine-learning algorithms operating on bond-formation and breaking energetics aid in rapid evaluation of highly reactive processes. We show that without any apriori chemical intuition, our automated framework recovers the most favorable reaction paths to form key SEI components, which were carefully identified over the past two decades. Thus, our data-driven approach and infrastructure show promise to accelerate the understanding of chemical reactivity in complex environments, in the aid of novel electrolyte design.

595758 Accelerating the Discovery of Novel Perovskites with Robots and Data

Abstract:

In this talk, I will describe our Robotic-Accelerated Perovskite Investigation and Discovery (RAPID) system. The first generation of RAPID uses inverse temperature crystallization (ITC) to grow halide perovskite single crystals for x-ray structure determination and bulk characterization using commercial liquid handling robots. The second generation extends this to antisolvent vapor diffusion experiments. Experiment plans for the syntheses are contributed remotely, by both human scientists and algorithms trained on the reaction data, facilitating by our ESCALATE (Experiment Specification, Capture and Laboratory Automation Technology) data management system. Incoming data collected by ESCALATE is used to automatically train machine learning models, evaluate model performance and feature influence, and quantify reproducibility. A live web dashboard communicates these insights to the scientist and management in visual form. I will conclude by describing case studies about new scientific insights extracted from the comprehensive RAPID dataset that have been enabled by this comprehensive dataset, and discuss ongoing deployments of ESCALATE to perovskite thin film and vapor diffusion synthesis experiments.

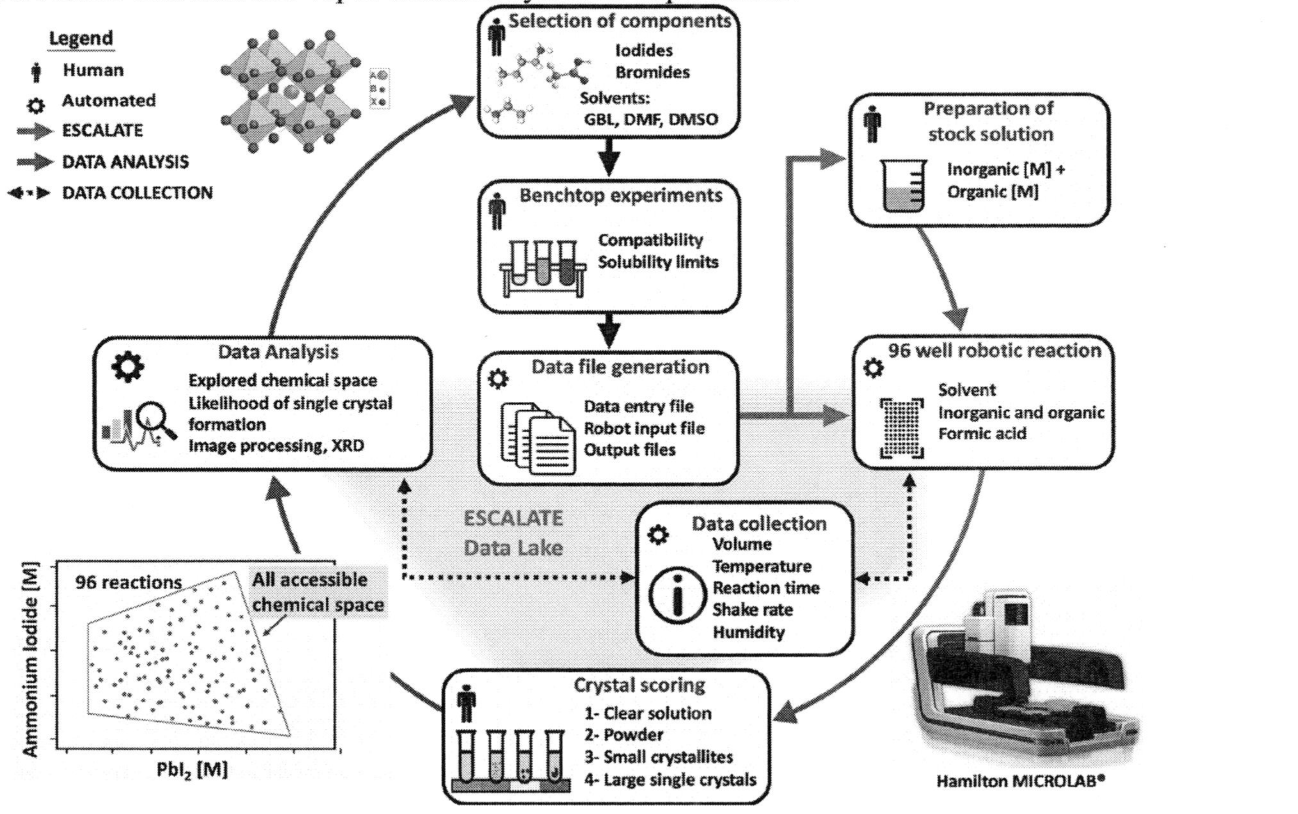

ESCALATE: Experiment Specification, Capture and Laboratory Automation Technology: A software pipeline for automated chemical experimentation and data management

598796 Streamlining Small Molecule Design and Synthesis

Abstract:

This talk will cover recent progress in the application of data science and machine learning techniques to problems in synthetic chemistry as they relate to molecular discovery. The typical discovery paradigm is an iterative process of designing candidate compounds, synthesizing those compounds, and testing their performance, where each repeat of this cycle can require weeks or months. Time and cost constraints may necessitate selecting compounds on the basis of what are perceived as fast to synthesize, rather than what are most informative to assay. Data science and statistical learning offer unprecedented opportunities to systematize and streamline the process by which new functional small molecules are designed and synthesized. I will describe how historical reaction data can be used to inform decision-making in small molecule pathway design, for both retrosynthesis and forward reaction prediction, and how these techniques can be integrated with *de novo* molecular design to triage candidate compounds for testing.

609606 Transitioning from ‘Traditional’ to Data-Driven 'wet' Laboratories: Growing-Pains and Future Outlook

Abstract:

The adoption of data-science methods within chemical engineering is primed to revolutionize materials discovery and to accelerate developments in molecular understanding. Yet, the movement is largely dominated by data and information originating from molecular modelling and simulation, or that is acquired from legacy sources (e.g. databases, literature mining) accumulating data from decades of careful experimental work. In order to advance data-driven materials developments well into the future, high-throughput experimentation (HTE) and automation throughout the complete laboratory workflow must also be developed and widely adopted to accelerate rates of experimental data production. In this talk we outline several examples and experiences showcasing how our research group is developing and adapting hardware and software infrastructure to accelerate the pace of molecular discovery in soft-matter systems for applications in health care, clean energy and materials synthesis. The talk will highlight recent research examples related to the implementation of HTE for electrolyte discovery and colloidal formulation/synthesis. We will also highlight significant challenges that have emerged from transitioning an established ‘wet-laboratory’ practice to HTE. These relate to adapting routine experimental methods to achieve HTE, developing new skills within the research workforce, the adoption of new data stewardship practices, needs for autonomous data sorting/classification, algorithms for automatic modeling and analysis and many others. Conversely, we will also highlight the numerous opportunities that emerge for enhancing virtual collaboration, enabling open data/hardware/software sharing, tackling challenging irreducible problems (e.g. optimization of complex formulations), and improving the outlook for the implementation of 'self-driving' laboratories.

612393 Accelerating Molecular Discovery with Machine Learning and AI

Abstract:

Deep learning is revolutionizing many areas of science and technology, particularly in natural language processing, speech recognition and computer vision. In this talk, we will provide an overview into latest developments of machine learning and AI methods and application to the problem of drug discovery and development at Isayev's Lab at CMU. We identify several areas where existing methods have the potential to accelerate pharmaceutical research and disrupt more traditional approaches. First we will present a deep learning model that approximate solution of Schrodinger equation. Focusing on parametrization for drug-like organic molecules and proteins, we have developed a single 'universal' model which is highly accurate compared to reference quantum mechanical calculations at speeds 10^6 faster. Second, we proposed a novel computational method for *de-novo* design of molecular compounds with desired biological and physical properties.

606261 Determination of Orientation and Conformation of Lysozyme at the Air-Water Interface Using an Integrated MD/SFG Approach

Abstract:

Proteins have markedly complex and diverse structures due to a high variability in size, shape, amphiphilicity, and charge needed to carry out their biological function. The interfacial behavior of adsorbed proteins govern properties such as foaming and stabilization, relevant to applications in pharmacy, biotechnology, and the food industry.[1-2] By studying proteins at the air water interface (AWI), which occurs naturally in many biological processes, we can explore how conformation and orientation of these complex biomolecules lead to self-assembly at these interfaces. To date, the interfacial structure of even one of the most widely studied proteins, lysozyme, remains unresolved.

In this talk we describe our approach to uncovering this interfacial structure by combining molecular dynamics simulation, vibrational sum frequency generation (SFG) spectroscopy, and spectral calculations to determine the conformation and orientation of lysozyme at the AWI. Computationally, two force fields are used to simulate lysozyme. With this approach we determine agreement in a single interfacial pose at high atomistic resolution. We validate the proposed structure of lysozyme by comparing signals of experimentally derived structures, to the spectra calculated from simulation, showing strong agreement by the pose predicted by MD. Lastly, we provide additional atomistic insight, discussing how pH may lead to an orientation change from head on to side-on at the interface, explaining previous discrepancies in the literature with regards to the thickness of the experimentally derived monolayers. This work provides a template for future studies of proteins at interfaces to make maximum use of integrated computational and experimental approaches.

[1] Dickinson, E. Proteins at Interfaces and in Emulsions Stability, Rheology and Interactions. *J. Chem. Soc. Faraday Trans.* **1998**, *94* (12), 1657–1669.

[2] McClements, D. J. Protein-Stabilized Emulsions. *Curr. Opin. Colloid Interface Sci.* **2004**, *9* (5), 305–313.

600717 Density Functional Theory Model for Adsorption-Induced Deformation of Materials with Non-Convex Pore Geometry

Abstract:

Adsorption of fluids in nanoporous media causes mechanical stresses and results in deformation [1]. This phenomenon is ubiquitous, and its magnitude depends on the pore size and geometry. Adsorption and adsorption-induced deformation is typically modeled in convex pores (cylindrical or spherical pores) or planar geometry (slit pores). However, many porous materials are composed of spherical grains, so that the pores are formed by the intergranular spaces between the convex surfaces, thus becoming non-convex. Here we present a first theoretical study of adsorption-induced deformation of materials with non-convex pores. We take the templated mesoporous carbons (CMK-3) which have an ordered hexagonal structure of parallel cylindrical rods [2], and model adsorption of nitrogen in this system using the classical density functional theory (DFT) within the local density approximation. Adsorbed fluid density profile in CMK-3 geometry can be reduced to two dimensions, but unlike slit or cylindrical pore model it cannot be made one dimensional. The fluid-fluid interaction parameters are fit to experimental nitrogen binodal, and reproduce other thermodynamic properties (such as surface tension and vapor pressure). The solid-fluid interaction parameters are fit to reproduce the nitrogen adsorption at 77 K on a reference carbon surface. Using the resulting parameters we predict the adsorption isotherms and strain isotherms for nitrogen adsorption in CMK-3 carbons. The shape of adsorption isotherm matches the shape of experimental isotherm. The predicted strain isotherms are qualitatively different from the strain isotherms in cylindrical pores [3]. We attribute this difference to formation of liquid bridges between the adjacent rods [4]. Our results suggest that adsorption-induced deformation in materials with non-convex pore walls cannot be predicted within the existing models. These results may shed some light on understanding adsorption-induced deformation of consolidated granular media.

References:

1. Gor, G. Y.; Huber, P.; Bernstein, N., 2017. Adsorption-induced deformation of nanoporous materials - a review. Applied Physics Reviews, 4(1), p.011303.
2. Jun, S.; Joo, S. H.; Ryoo, R.; Kruk, M.; Jaroniec, M.; Liu, Z.; Ohsuna, T.; Terasaki, O., 2000. Synthesis of new, nanoporous carbon with hexagonally ordered mesostructure. Journal of the American Chemical Society, 122(43), pp.10712-10713.
3. Gor, G. Y.; Neimark, A. V., 2011. Adsorption-induced deformation of mesoporous solids: Macroscopic approach and density functional theory. Langmuir, 27(11), pp.6926-6931.
4. Ludescher, L.; Braxmeier, S.; Balzer, C.; Reichenauer, G.; Putz, F.; Hüsing, N.; Gor, G. Y.; Paris, O., 2020. Capillary Bridge Formation between Hexagonally Ordered Carbon Nanorods, Adsorption, DOI: 10.1007/s10450-020-00215-6

604204 Optimal Packing of CO at High Coverage on Pt(100) and Pt(111) Surfaces

Abstract:

High coverage structures for CO on Pt(111) and Pt(100) surfaces are studied by density functional theory modeling and compared to high pressure scanning tunneling microscopy experiments. Semilocal exchange correlation functionals are known to provide incorrect adsorption site and overestimated adsorption energy for CO on Pt. We develop a simple first-principles correction for the adsorption energy of CO on Pt(111) and Pt(100) using the bond length of adsorbed CO as a descriptor. The developed correction which decreases in the order top > bridge > hollow site is used to derive the surface stability diagram for CO adsorbed on 111 and 100 facets of Pt showing the thermodynamically stable CO configurations on the surface as a function of temperature and pressure. High coverage (θ>0.5) configurations of CO on Pt(111) lead to formation of superimposed hexagonal/quasi-hexagonal lattice of CO on hexagonal Pt(111) layer from a systematic exploration of such structures. Non-hexagonal CO structures seen in vacuum and low temperature conditions - c($\sqrt{3}$x3)rect-4CO, c($\sqrt{3}$x5)rect-3CO, and c($\sqrt{3}$x7)rect-5CO are however only 5 meV/Å^2 less stable than the hexagonal lattice at similar coverage. For Pt(100) at θ>0.75, the CO molecules adopt a one dimensional coincidence lattice and we observe the formation of (nx2) unit cells (n=4,6,8) with (2n-2) CO molecules in each cell on top/quasi-top and bridge/quasi-bridge positions creating a skewed hexagonal lattice to reduce CO-CO repulsion with increasing coverage. The computational results agree with the available experimental observations for Pt(111) and Pt(100). The integrated theoretical simulation and experimental observation provide structural data for the study of catalytic reactivity on Pt surfaces in reactions involving high CO pressures and suggest an approach for understanding roles of CO molecules on other metal catalyst surfaces in high pressure CO and during catalysis CO participates.

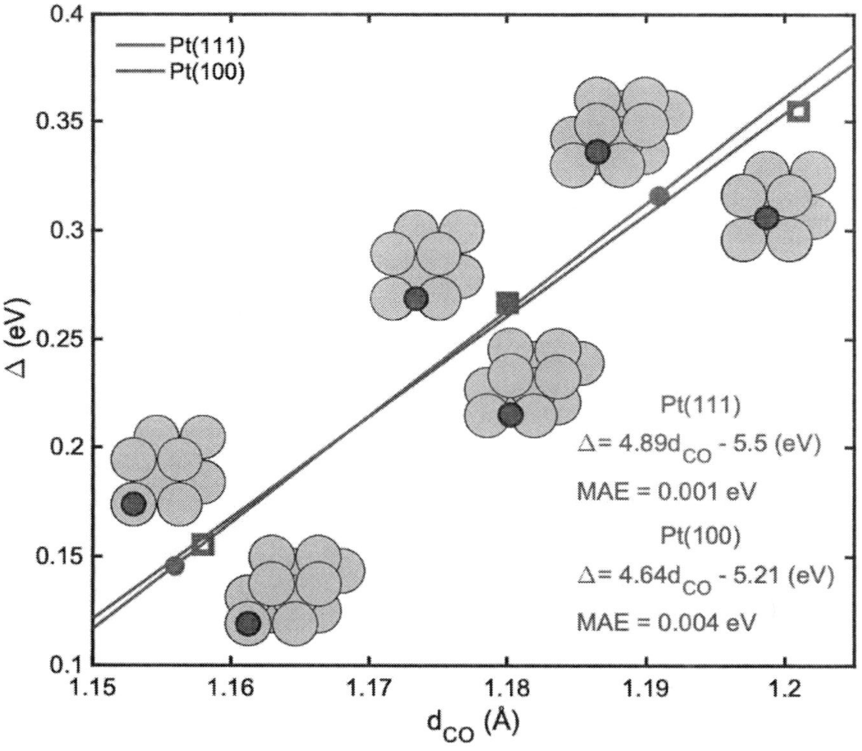

609577 Quantitatively Accurate Theory to Predict Adsorbed Configurations of Surfactants at Polar Surfaces

Abstract:

Surfactant molecules adsorb on polar surfaces in various morphologies, such as planar, spherical and cylindrical. So far, there is no encompassing theoretical framework that related molecular properties to the adsorbed morphologies. We have developed a theoretical model that is able to predict adsorbed morphologies of surfactant molecules on polar surfaces given their geometry and strength of interactions with the surface. We show, via coarse-grained simulations, that our theoretical model is quantitatively accurate in predicting the adsorbed configurations. Specifically, our model accurately predicts that linear-shaped surfactants adsorb either as a self-assembled monolayer or as stripes on the surfaces, depending on the relative interaction strengths of the polar head group and the alkyl tails with the surface. Our model correctly predicts the formation of different cylindrical and spherical configurations of the adsorbed asymmetrically shaped surfactant molecules.

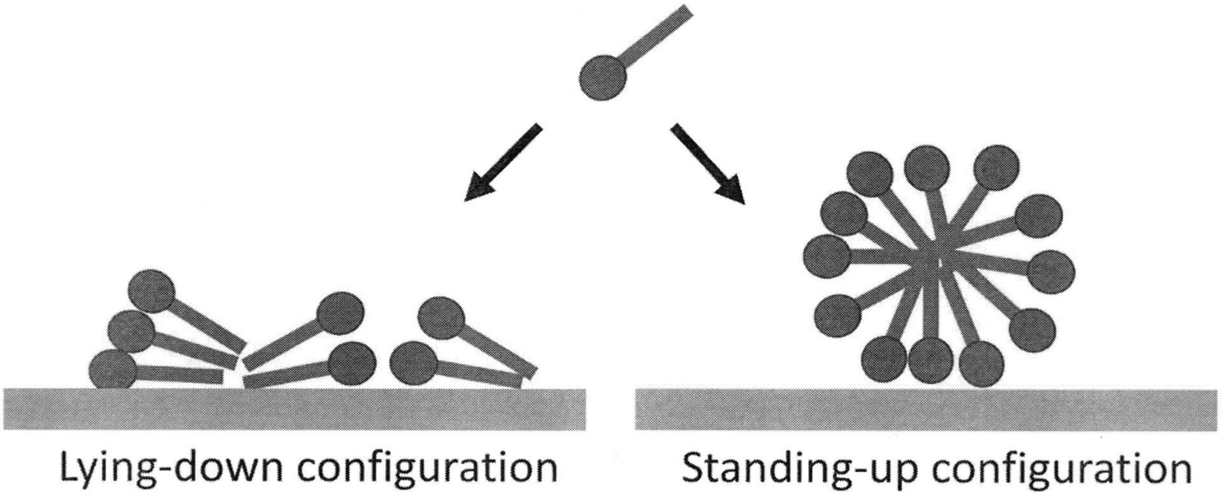

601439 2D Phase Transitions of Fluid Adsorbed in Cylindrical Mesopores: Influence of Adsorbent-Adsorbate Interactions

Abstract:

Recently, molecular simulations of methane adsorption in mesoporous cylindrical silica identified a two-dimensional phase transition associated with non-trivial densification of the adsorbed liquid, a high-density local maximum in the isosteric heat of adsorption, and hexagonal rearrangement of annular layers of adsorbed fluid [1]. This observed rearrangement of adsorbed methane was proposed as a plausible explanation for the densification of methane adsorbed in MCM-41 and SBA-15 that was observed in small-angle neutron scattering (SANS) experiments [2]. The proposed phase transition was, however, unique to the methane-silica adsorption systems, not being observed in, for example, nitrogen-silica, argon-silica, or argon-carbon systems. To better understand this unique phase transition and why it occurs for the methane-silica systems in contrast to others, we perform additional Monte Carlo molecular simulations to identify key characteristics of the methane-silica system that give rise to its appearance. Specifically, we measure adsorption isotherms, heats of adsorption, and bond-orientational order parameters for the benchmark systems that do not exhibit such phase transitions, and then smoothly transform those benchmark systems into the methane-silica system to clearly identify the boundaries in adsorbent-adsorbate parameter space that trigger this type of phase transition. Lastly, we discuss engineering consequences of the observed phase transition and how it may impact other thermophysical properties of the adsorbed fluid.

[1] D.W. Siderius, W.P. Krekelberg, W-S. Chiang, V.K. Shen, and Y. Liu, *Langmuir*, 33:14252 (2017).
[2] W-S. Chiang et al., *Langmuir*, 32:8849 (2016).

597618 Impact of Solvated Water Layer on Adsorption of Selenium Oxo-Anion on <012> Hematite Surface

Abstract:

Selenium is highly soluble in water as selenate and selenite and is toxic to human and aquatic life even at very low concentration (TWA 0.2 mg/m^3). Increased usage of selenium in the electronic and glass industry threatens to further increase the concentrations of these species in our water resources. Adsorption is a promising remediation strategy, particularly in low population densities area and communities with self-run wells, because of its potentially lower cost than the construction of chemical treatment plants. The adsorption of oxo-anions on existing sorbents is , however, not selective; this is exacerbated by the fact that Selenium oxoanions occur at much lower concentrations than similar S ions. This leads to increased financial burden for water treatment. Recent adsorption experiments of selenium oxoanion of different hematite surface found that selenate and selenite have higher selectivity on [012] and [110] hematite facet respectively. Here we use first principle calculations to study the facet dependent behavior of hematite based adsorption. Adsorption energies are calculated for varying oxo-anion configuration including inner and outer sphere complexes on [012] surface. In this work, we delineate the most important effects controlling adsorption, including surface configuration, pH, solvation, etc. We find the solvation configuration directed by the surface controls the adsorption behavior. In addition to shedding light on this particular adsorption species, this work highlights the criticality of correctly considering solvation, and outer sphere adsorption, in first principles analysis of oxoanion adsorption.

600557 Microkinetic Modeling Strategy Considering Topologic Aspect of a Cu Catalytic Surface Based on Kinetic Monte-Carlo

Abstract:

Introduction

It is obvious that surface reactions occur between "neighboring" reactants adsorbed on the surface, but conventional microkinetic modeling has not been conducted majorly in this respect. Traditional microkinetic modeling methods, also in our previous study too, consist only of surface coverage, which is the ratio how much a reactant is adsorbed on the surface, -oriented rate equations. Moreover, the reaction of the gaseous reactants is usually calculated by equations based on chemical equilibrium, and the surface reactions are usually calculated by equations based on reaction kinetics. We thought that this inconsistency should be improved so that we used kinetic Monte-Carlo method on microkinetic modeling.

Incorporating kMC simulation into a microkinetic model provides three advantages:

1. The model can take it account into that the topologic effects of surface reactions.
2. All reactions, including gas phase reactions, can be modeled based on reaction kinetics not on equilibriums.
3. The diffusion reactions on the catalytic surface which is not possible with conventional microkinetic plug-flow reactor (PFR) 1D modeling, can be considered.

On the other hand, there are two major disadvantages of the method of simulating the catalytic surface with kMC:

1. The computational load is greatly affected by the size of the catalytic surface.
2. The calculation time is much long compared to the general microkinetic model, which is disadvantageous for parameter estimation.

In order to solve the first problem, this study used periodic boundary unit cell structure to overcome the finite size effect problem.[2,3] Also, to solve the second problem, we derived the correlation between kinetic constants in two different scales of which are molecular level based on kMC, -number oriented form and continuum level based on general microkinetic model, -concentration oriented form. To drive that correlation, we used Gaussian Process (GP) based on Bayesian Optimization using Deep Neural Network (DNN) and performed parameter estimation using microkinetic model using the rate constant oriented from kMC simulations. [4]

Materials and Methods

Our system is Cu catalytic system consists of syngas which is a composition of CH4, CO and CO2 gases. The Cu catalytic surface of our system is based on Cu (111), and the lattice structures of unit cell are varied with 10-by-10, 20-by-20 and 50-by-50. Experimental data of adsorption / desorption reactions on Cu surface at equilibrium will be used to estimate kMC parameter, especially on desorption kinetics.

The reactor type is PFR and it is modeled by a series of Continuous Stirred Tank Reactors (CSTRs). With no assumption of rate determinate step (RDS), the model will be a microkinetic model with total

26 reactions which are taken from the work of C. V. Ovesen. *et al.* consist with five adsorption/desorption reactions and eight reversible, forward and backward surface reactions. [5] Kinetic constants, pre-exponential factors, activation energies and binding energies are taken from our previous study and Ovesen, C.V., *et al.*'s work which are tuned with experimental data. [1,5]

The tools we used are total two. One for DNN is TensorFlow® library of python and another for kMC simulation and optimization of microkinetic model with kinetics is MATLAB®.

Significance

In point of science, it has its novelty at taking topological reaction into account microkinetic modeling. In addition, we can present the adequate strategy for modeling a desorption kinetics from continuum scale experimental data which has been unclear yet.

601219 Application of Derjaguin-Broekhoff-De Boer Theory and Molecular Simulation for Calculation of Solvation Pressure in Spherical Mesopores

Abstract:

Adsorption of fluids in porous structure takes place due to the intermolecular interactions between solid surface and adsorbate. During this process, the adsorbate confined in pore exhibits high "solvation" pressure on the pore walls and causes the deformation of porous materials [1]. The high pressure in the pores have additional effects other than mechanical strains, in particular, some of the high-pressure chemical reactions take place in the pores at the moderate gas pressures [2,3], or the compressibility of fluids in pores changes [4]. Here we focus on solvation pressure in model silica pores of spherical shape induced by the adsorption of nitrogen.

We use the thermodynamic model for calculating solvation pressure from adsorption isotherms, applied before for various microporous materials [5,6] and mesoporous materials with cylindrical pores [7]. We apply it for spherical mesopores using the adsorption isotherms predicted from the macroscopic Derjaguin-Broekhoff-de Boer (DBdB) theory [8-10] and grand canonical molecular simulations. We show that both approaches give consistent results, which are also typical for the solvation pressure in mesoporous materials, yet noticeably differ from the results for cylindrical pore geometry [7]. Furthermore, we show that the dependence of the solvation pressure on the reciprocal pore size can be used for the calculation of the solid-liquid surface energy. Finally, we relate the calculated solvation pressures to the compressibility of liquid nitrogen in spherical confinement and found that the departure of the compressibility from the bulk value is consistent with the calculated values of solvation pressures.

[1] Gor, G. Y., Huber, P., & Bernstein, N. (2017). Appl. Phys. Rev., 4(1), 011303.
[2] Kaneko, K., Fukuzaki, N., Kakei, K., Suzuki, T., & Ozeki, S. (1989). Langmuir, 5(4), 960.
[3] Byl, O., Kondratyuk, P., & Yates, J. T. (2003). J. Phys. Chem. B, 107(18), 4277.
[4] Gor, G. Y., Siderius, D. W., Shen, V. K., & Bernstein, N. (2016). J. Chem. Phys., 145(16), 164505.
[5] Ravikovitch, P. I. & Neimark, A. V. (2006). Langmuir, 22(26), 10864.
[6] Kowalczyk, P., Ciach, A., & Neimark, A. V. (2008). Langmuir, 24(13), 6603.
[7] Gor, G. Y. & Neimark, A. V. (2010). Langmuir, 26(16), 13021.
[8] Derjaguin, B. V. (1940). Acta Physicochim. URSS, 12, 181.
[9] Broekhoff, J. C. P. & De Boer, J. H. (1967). J. Catal., 9(1), 8.
[10] Broekhoff, J. C. P. & De Boer, J. H. (1968). J. Catal., 10(4), 153.

601991 The Digital Transformation of the Chemicals Industry - and How to Prepare

Abstract:

Over the course of the next 20 years, the chemicals industry will see more change than it has in the last 50, in large part due to digital technologies. Digital transformation is affecting business processes from R&D, to manufacturing, to supply chain, to sales and go-to-market -- and the data from all these digital efforts is tying those diverse functions to optimize the chemical enterprise and create new business models and sources of value. This presentation looks at how the industry will change through 2040 and examines what the new sources of competitive advantage will be for both companies and individuals.

613114 Driving the Value from Digitalization

Abstract:

Many opportunities exist to utilize the technologies of Industry 4.0 and digitalization within the process industries, to improve reliability, productivity, profitability, and safety. However, organizations may become enamored with the new tool or capability, looking for problems to fit the technology rather than focusing on the true business needs. Only by being technology agnostic and using a strategy that places balance sheet impact and long term competitive advantage as selection criteria to the myriad of solutions available in the market will an organization focus on the proper challenges. This presentation will examine potential pitfalls to technology selection and implementation.

613028 Combining AI with Chemical Engineering First Principles Will Take Process Performance, Safety and Sustainability to New Levels

Abstract:

Digitalization is now in the spotlight in terms of creating improved agility, resilience, and ability to operate process industry assets remotely and safely.

Sensors, connectivity, and cloud computing are making process data more available for use in optimization and prediction. High performance, edge computing, and new algorithms are enabling placing process models online for monitoring, optimization, and operational advice.
Most process and chemical engineers in industry have observed the influx of data science and data scientists into their organizations. This session will cover an area called "hybrid modeling." This combines and embeds AI and machine learning within first-principles rigorous models. You may think of this as putting chemical engineering guardrails around machine learning. Rigorous, first principles models, brings to bear decades of chemical engineering knowledge. Machine learning and AI brings to bear ability to rapidly analyze, and progressively learn from empirical operating data. When combined, it is now possible to achieve a new level of modeling that will further leverage both types of analysis. We have been validating this breakthrough combination of AI and chemical engineering with over 100 companies in industry, and will cover several high value and immediately practical use cases during this session.

601992 Designing Materials and Processes in the Digital World

Abstract:

3M has been known as a materials innovation company for many years with thousands of new products launched annually, many of those enabling and empowering everyday products. This innovation model is based on leveraging deep technologies into novel combinations to develop product platforms.

This talk will address the challenges and progress in transforming this culture and enabling it with digital tools created internally and with outside partners to accelerate the speed of materials innovation dramatically.

AUTHOR INDEX